Thomas Lechford

Plain Dealing

Thomas Lechford

Plain Dealing

ISBN/EAN: 9783337037147

Printed in Europe, USA, Canada, Australia, Japan

Cover: Foto ©berggeist007 / pixelio.de

More available books at **www.hansebooks.com**

PLAIN DEALING

OR

News from New England

BY

THOMAS LECHFORD

WITH AN INTRODUCTION AND NOTES

BY

J. HAMMOND TRUMBULL

Boston
J. K. WIGGIN & WM. PARSONS LUNT
M DCCC LXVII

TO

GEORGE BRINLEY, ESQ.,

OF HARTFORD,

A SUCCESSFUL COLLECTOR AND A CAREFUL READER OF ALL THAT ILLUSTRATES
THE EARLY HISTORY OF NEW ENGLAND,

WHATEVER THERE MAY BE OF VALUE IN

𝔗his 𝔈dition of "𝔓lain 𝔇ealing,"

THE PREPARATION OF WHICH WAS UNDERTAKEN ON HIS SUGGESTION,

IS,

IN SLIGHT ACKNOWLEDGMENT OF MANY OBLIGATIONS,

DEDICATED

BY HIS FRIEND,

THE EDITOR.

INTRODUCTION

BY THE EDITOR.

IN the year 1858, the late Samuel Jennison, Esq., of Worcester, — for many years an officer of the American Antiquarian Society, and well known as a diligent and successful student of the history of New England, — invited me to examine a manuscript volume of which he had become the possessor. Of this volume, its authorship and contents, Mr. Jennison wrote the following account: —

"It is now some time since a friend, knowing that I had some curiosity in relation to matters of antiquity, and thinking I might find something to gratify it in a small folio, in manuscript, in broken but venerable binding, which was then in his possession, placed the same in my hands. He did not know the writer or the contents; for the style of penmanship was that of more than two centuries ago, and although not unusually indistinct for its kind, could not easily be read by one unaccustomed to the chirography of the time. It proved, on inspection, to be a journal-book kept by THOMAS LECHFORD, whose claim to the reputation of having been the *first Boston lawyer* is, I believe, unquestioned. It contains a record of the business transactions in which he was, from day to day, engaged, commencing with his settlement in Boston, and continued until his return to England; embracing many facts of historical and genealogical interest. I have awaited a season of leisure and relief from other occupations to transcribe and prepare it for publication."

It is much to be regretted that the work of transcription and preparation, commenced by Mr. Jennison, was laid aside before being completed, and the public thereby deprived of the valuable

INTRODUCTION.

illustration such a volume must have received in the hands of so competent an editor.

In this Journal, Lechford had made numerous entries in short-hand, some of which are of considerable length, and one occupies an entire page. It was for the purpose of asking assistance in deciphering these, that Mr. Jennison first submitted the volume to my inspection. I recognized the characters employed, as belonging to a system with which I had previously become tolerably familiar, and promised to furnish the desired translations whenever I could find leisure for the work. Mr. Jennison died (March 11, 1860) before this promise could be redeemed. Until the spring of 1865, I found it impossible to devote the time requisite for the study of the cipher, and for a more thorough examination of the volume. On application to Samuel Jennison, Esq., of Boston, into whose possession the manuscript had come by the decease of his father, he not only most obligingly consented to intrust it to my hands for so long a time as should be necessary for deciphering the short-hand, but subsequently, with a liberality for which I am glad to have this opportunity of acknowledging my obligations, authorized me to publish, in a limited edition, the entire manuscript, and materially lightened the labor of preparation by permitting me to make use of an abstract of the volume and an index, which his father had made.

The first instalment of this publication was nearly ready for the printer when Lechford's *Plain Dealing* was announced for re-impression in the "Library of New-England History." I consented to become the editor of this volume, because it seemed desirable that it should receive the benefit of whatever new material the author's journal and manuscript letters might supply for its illustration, and because much of this material might be more advantageously employed in notes to *Plain Dealing* than in a separate publication. Meanwhile, the preparation of the Journal for the press has been suspended. But the work is already resumed, and a volume will shortly be published com-

. prising Lechford's entries of business transactions, copies or abstracts of instruments drawn by him, and letters to his correspondents in New and Old England, between June 27, 1638, and Dec. 31, 1639. For that volume, the ensuing sketch of the little that is known of Lechford's personal history, and estimate of the man and his book, was originally prepared. The prior announcement of *Plain Dealing*, by the publishers, and its necessary precedence in the series, have compelled me to employ the same materials, in almost the same form, by way of introduction to two separate but nearly connected publications.

Of the birth and parentage of Thomas Lechford, or of his early life, I have no certain knowledge. His surname is that of a family, which, at about the middle of the sixteenth century, became seated at Leigh, near Reigate, in the county of Surrey, where Henry Lechford, great-grandson of a Thomas Lechford who lived in the reign of Edward IV. (1461–1482), bought the manors of Shellwood and Charlwood, with other estates, This Henry, dying, Sept. 27, 1567, left a son Richard, born in 1547, who was knighted. Sir Richard Lechford was twice married; first, to Ann, daughter of George Lusher, by whom he had two sons, John and *Thomas;* and, secondly, to Eleanor, daughter of William Morgan, of Chilworth, Esq. Henry, a son of the second marriage, died in 1606, before his father, but left a son Richard (born, about December, 1594), who inherited the estates of his grandfather on the death of the latter, July 10, 1611. John and Thomas, above named, sons of Sir Richard by his first wife, were living in 1606, when they are named in a deed of settlement by their father on his second wife and her children.* Their nephew, Richard Lechford, was knighted by James I. Early in the reign of Charles I., he was enrolled in the band of "Gentlemen Pensioners," who constituted the king's body-guard. Like many other courtiers of his day, he became a Roman Catholic, and found his new religion no bar to royal favor, notwithstanding the

* Manning and Bray's *History of Surrey*, ii. 181, 184–5, 188.

unabated severity of the Laws against "popish recusants." His eldest daughter, Letitia ("*alias* Bridget," as she is named in the record), remained a Protestant, and, about 1633, was confirmed in the Church of England, to the great displeasure of her father.* Not long afterwards, while Sir Richard was in attendance upon the king in his journey to Scotland, this daughter Letitia and a younger sister Mary, who had been placed under the care of an aunt living near London, were detained by warrant from the High Commission when about to sail from England for some foreign port. Their father alleged that they had embarked without his knowledge, and were attempting to escape from his authority; but another and more probable version of the story is given by a well-informed writer (the Rev. George Gerrard, the gossipping London correspondent of the Earl of Strafford), in a letter dated May 1, 1634: †—

"Sir Richard Lashford,‡ a penfioner in ordinary, was fending two of his daughters to the nunneries beyond the feas; being to take fhipping in fome of the Kentifh ports, they were ftopp'd and fent back to London. My Lord's Grace of Canterbury [Laud] being made acquainted with it, fent for the father, who offered to give caution that they fhould not go out of England; but my Lord afked him, whether he would engage himfelf that they fhould conform themfelves to the religion of the Church of England, which he refufed. He afked then of him, of what religion he was? He faid, A Romifh Catholick, and but lately converted. He offered him both the Oaths, which peremptorily he refufed. The Archbifhop then told him, he was not a fit fervant to be of the King's principal Guard, that would not take the oath of allegiance unto his Majefty. Since he hath been brought before the Lords, abfolutely put out of his place, and another fworn into it."

---

* *Calendar of Brit. State Papers* (Domest. Ser., Charles I.), 1633-4, pp. 23, 536, 348, 581.

† Strafford's *Letters and Dispat.*, i. 242.

‡ So the name appears to have been generally pronounced, and was occasionally written. Evelyn (*Diary*, ii. 56, Bohn's edit.) mentions, under date of Sept. 13, 1670, going "to vifit Sir Richard Lashford, [his] kinfman." Elfewhere, we find the same name written *Leecheforde*. See note (†) on the next page.

INTRODUCTION.

A few weeks afterwards, the same correspondent wrote : *—

"The Penfioner, Sir Richard Lafhford, was again called before the Lords, when the oath of allegiance was again offered to him, but he utterly refufed it. So order was given to the Attorney to indite him in the King's Bench of a *Premunire;* but being brought thither, he took it before the Judges, which if he had done before, 'tis likely he had not been put from his Penfioner's place."

In other words, the influence of the court upon the judges, or a "letter of grace and protection," such as the king, about this time, was accustomed to grant to his courtiers who were papists, would have stayed proceedings against him for recusancy.

In November, 1634, Sir Richard sold the manor of Shellwood, and other estates in Surrey, and subsequently resided at or near Dorking (in the same county), where he died, Sept. 14, 1671.†

The recurrence of the name of *Thomas* in several generations of the Lechfords of Shellwood; the fact that the surname was by no means common, and does not appear to have been represented in England by any other family than this, of the rank of gentry; with other considerations which it is unnecessary to mention here,—render it highly probable, in fact nearly certain, that the author of *Plain Dealing* was of this stock, and nearly related to the last-mentioned Sir Richard Lechford, Knt., 1634.

In the address "To the Reader," Thomas Lechford describes himself as "a student or practiser at law." An entry in his journal shows that he had been a member of Clement's Inn *before* he came to New England; and he resumed his residence there after his return to London, in 1641, as the title-page of *Plain Dealing* informs us. In an order of the General Court of Massachusetts,

* Strafford's *Letters and Dispatches,* i. 261.
† Manning & Bray, i. 586. The arms confirmed to "Sir Richard Leechforde of Shelwood," Nov. 22, 1605, by W. Segar, Garter, are thus blazoned : *Sable,* a chevron betw. three leopards' heads, *argent.* Crest, on a wreath of the same colors, a unicorn's head erased, *argent,* maned, bearded, and horned *or,* bearing on the same a serpent proper. Howard's *Miscel. Geneal. et Herald.* (Oct.1866), p. 54.

INTRODUCTION.

made in 1647, he is described as "an ordinary *solicitor* in England."* It does not appear that he was ever called to the bar. The Inns of Chancery, of which Clement's was one, were so called "probably because they were appropriated to such clerks as chiefly studied the *forming of writs*, which was the province of the cursitors, who are officers of chancery, and such as belong to the courts of common pleas and king's bench." † In Stowe's time, they were "chiefly filled with attorneys, solicitors, and clerks." By an order of the judges, April 15, 1630, "attorneys and solicitors, which are but ministerial persons of an inferior nature," though permitted to occupy chambers in the inns of chancery, were excluded from the inns of court, and consequently from a call to the bar. ‡ In his defence before the court of magistrates at Boston, in December, 1640, § Lechford said of himself: "I am no pleader, by nature; oratory I have little, . . . and if I had never so expert a faculty that way, I should not now use it, . . . and as for the other part of pleading which consisteth in *chirography*, ‖ *wherein I had some little skill*, I do not desire to use any of that," &c.

When Hugh Peters was lecturer in St. Sepulchre, in London, — before the persecution of Laud drove him to Rotterdam, in 1629 or 1630, — Lechford was one of his hearers, and "hung upon his ministry," as he expresses it in a letter to Peters, written in 1639.¶ Some years later, he was in Ireland, with Sir Thomas Wentworth (afterwards Earl of Strafford), then lord deputy. In what capacity he went, or how long he remained

---

* *Mass. Col. Records*, ii. 206.
† Herbert's *Inns of Court and Chancery*, 169.
‡ Dugdale's *Origines Judicales*, 320.
§ See after, page xxxiii.; and note 256, on page 157 of this volume.
‖ This word appears to be used here in its more modern sense, for the business of a draughtsman and scrivener. In the old law, a chirographer signified "him in the Common Pleas office (*in Communi Banco*) that ingroffeth Fines in that Court acknowledged . . . and that writeth and delivereth the *Indentures* of them unto the parties" (Minsheu, 1627); and a chirograph was a bill, bond, or deed-indented, written in the maker's own hand.
¶ Copied, in short-hand, in his *Journal*.

there, does not appear.* In 1640, when he contemplated departure from New England, he wrote to one of his correspondents, that he was desirous to return to Ireland, "there to follow his old profession, where he had some hope of friendship." †

In the address "To the Reader," of *Plain Dealing*, he alludes to the fact, "well knowne unto many, that heretofore he suffered imprisonment, and a kind of banishment, . . . for some acts construed to oppose, and as tending to subvert Episcopacie, and the settled Ecclesiastical government of England." His offence, as we learn from a couple of lines in Mr. Cotton's *Way of Congregational Churches cleared*, was his "witnessing against the Bishops, in soliciting the cause of Mr. Prynne." In the judgment of Laud and of the High Commission, his crime could hardly have been greater, or have merited more severe punishment. Prynne, a barrister of Lincoln's Inn, had drawn upon himself the vengeance of the archbishop, by the publication, in 1633, of *Histriomastix*. He was indicted in the Star Chamber, found guilty of a libel, and condemned to a barbarous punishment, to be followed by imprisonment for life, for the crime of railing "not only against Stage Plays . . . but farther in particular against Hunting, Publique Festivals, Christmas-keeping, Bonfires, and Maypoles," &c. ‡ His *real* offence (as Hume suggests) was, probably, that he had, "in plain terms, blamed the hierarchy, the innovations in religious worship, and the new superstitions introduced by Laud." Four years afterwards, a renewal of this offence called for a yet more vindictive prosecution in the same court. On the 14th of June, 1637, he, with Henry Burton, bachelor of divinity, and John Bastwick, a physician, was tried and convicted of

---

* Wentworth was appointed lord deputy in January, 1632, but did not go to Ireland until July, 1633. In June, 1636, he came to London, remained about six months in England, and returned to Dublin in November. He was not again in London until September, 1639. In December, 1639, he was created Earl of Strafford and Lord Lieutenant of Ireland. — Strafford's *Letters and Dispatches*, i. 63, 84; ii. 430, 431; Nalson's *Collection*, i. 280.

† Letter, without address, dated July 28, 1640, copied, in short-hand, in his *Journal*, p. 159.

‡ *Rushworth*, ii. 220.

"writing and publishing seditious, schismatical and libellous books against the hierarchy of the Church." They were sentenced to lose their ears in the pillory, to be fined £5000 each to the king, to perpetual imprisonment in three remote places of the kingdom; and Prynne to be branded on both cheeks with the letters S. L., for a "Seditious Libeller." This barbarous sentence was executed in the palace-yard at Westminster, June 30; "a spectacle no less strange than sad, to see three of several professions, the noblest in the kingdom, Divinity, Law and Physick, exposed at one time to such an ignominious punishment, and condemned to it by Protestant magistrates, for such tenets in religion as the greatest part of Protestants in England held, and all the reformed churches in Europe maintained." * Immediately after summons was issued for Prynne's appearance before the court, he was shut up close prisoner, refused the use of pen, ink, or paper, and not permitted to consult counsel until very shortly before his trial. In his speech to the court he said: "I was deserted of all means by which I should have drawn my answer. ... I had neither pen, ink, nor servant to do any thing for me; for my servant was then also close prisoner, under a pursuivant's hands." All who rendered the slightest service to Prynne or his fellow-offenders fell under condemnation. "One Gardener," a scrivener or clerk, who wrote from Prynne's dictation a petition to his judges, was apprehended, subjected to fourteen days' imprisonment, and compelled to give a bond for appearance when called. His counsel, Holt and Tomlyns, did not dare to subscribe his answer, after it was drawn and engrossed. After the execution of his sentence, some of his friends visited him in Chester, on his way to his prison at Caernarvon. Those who had so offended were summoned before the Privy Council, cited into the High Commission at York, imprisoned and fined, and enjoined to make a public recantation.† It is not surprising that Lechford, for "soliciting" in Prynne's cause or otherwise assist-

---

* May's *Hist. of the Parl.*, b. i. ch. 7.   † Hargrave's *State Trials*, i. 482, 501.

ing his defence, should have been severely dealt with. Of his punishment we know no more than he himself has told us,—that he "suffered imprisonment and a kind of banishment."

Lechford landed in Boston one year and thirteen days after Prynne's trial in the Star Chamber. Four years and five months after the trial (Nov. 16, 1641), he dated his "Quæres about Church Government" from his chambers in Clement's Inn, and, on the first page of *Plain Dealing*, speaks of "having been forth of his native country *almost* for the space of four years last past." The inference, from comparison of these dates, seems to be, that he left England in the autumn or winter of 1637, but did not then sail directly for Boston. His imprisonment could not have been of many months' duration.

In the letter to Hugh Peters,* before cited, he writes : —

"Being thrown out of my ſtation in England ... I forſook preferment in a Prince's court that was offered to me, who of Chriſtian princes is the chiefe for godlineſs (as I was aſſured), Georgius Ragotzki, Prince of Tranſylvania and Lord of Lower Hungary, ſucceſſor to Bethlem Gabor.† Likewiſe the Lords of Providence‡ offered me place of pre-

---

* Deciphered from the short-hand copy in the *Journal*, p. 30.

† George, son of Sigismund Rakoczy, or Ragotzki, the representative of a noble family distinguished for many generations in the annals of Transylvania, was chosen prince (vaivode) of that province in 1631. He married a daughter of Stephen (brother of Gabor) Bethlem. As a champion of the Protestant cause in Hungary and Bohemia against the Jesuits and their tool, the Emperor Ferdinand II., and afterwards as the ally of Gustavus Adolphus, his name was held in high honor among the Protestants of Western Europe. Hoffmann (*Lexicon Univ.*) calls him "Princeps pacificus et egregius."

‡ The Earl of Warwick, Henry Rich Earl of Holland, Lord Say and Sele, Lord Brooke, Sir Benjamin Rudyerd, and Sir Nathaniel Rich, were among the "Adventurers for the Plantation of the Islands of Providence, Henrietta, and the adjacent Islands" (the Bahamas), incorporated by patent of Dec. 4, 1630. In 1636 and 1637, the privileges of the company were enlarged, and they were encouraged to make liberal advances for promoting the growth of the plantation and fortifying Providence Island against the Spaniards. In February, 1638, the Earl of Warwick, Lord Say and Sele, and Lord Brooke declared their intention of going themselves to the Island ; and a considerable number of planters and servants, with a supply of vessels, were to be sent thither

ferment with them, which I will not name. Hither I have come, and, the Lord knows my heart! fain would I join with your Churches," &c.

I have not been able to discover the time or place of Lechford's embarkation for New England, nor in which of the twenty ships which brought three thousand passengers to Massachusetts in the summer of 1638,* he came. His journal begins with the date of his arrival: —

"Bofton in New-England, 27º 4! the day of my landing — 1638."

From some allusions in his letters, especially a reference to conversation "on ship-board," I infer that he came fellow-passenger with Mr. Edmund Browne, afterwards minister of Sudbury, and, perhaps, with Emanuel Downing,† the brother-in-law of Governor Winthrop.

From succeeding pages of his journal, we gather some — but scanty and unsatisfactory — knowledge of his domestic relations. His wife is mentioned, in 1639 and afterwards; and, as no evidence has been discovered of his marriage on this side of the water, we infer that she accompanied him from England; but he nowhere gives any information of her family, nor even introduces her Christian name. In July, 1640,‡ he writes: "I have not yet here an house of my owne to put my head in, or any stock going." He lived in a house, or part of a house, hired of Nathaniel Micklethwaite of Boston, who was, I think, the agent or factor in New England of Richard Hutchinson of London, and perhaps of Edward and William Hutchinson after their removal to Rhode Island.

in advance of the coming of the Lords. Great inducements were offered to planters, and strenuous efforts made to divert emigration from New England to Providence. Among others thus solicited were the Rev. Charles Chauncy, the Rev. Ezekiel Rogers, and Capt. John Underhill. — Sainsbury's *Calendar*, 123, 248, 262, 267. See *Plain Dealing*, p. 48, and note 198.

* *Winthrop*, i. 268.

† Yet I find elsewhere no earlier mention of Downing's arrival than that in the records of the Court of September 6, 1638. (*Mass Records*, i. 236.) Mr. Savage had, apparently, overlooked that reference, when he wrote the note to *Winthrop*, i. 274.

‡ *Plain Dealing*, p. 69.

It appears that he paid his rent, until August, 1639, to Samuel Hutchinson, and subsequently to Mr. Micklethwaite, whose signature appears, on a page of the journal, to the lease of "the chamber etc.," at £5 per year, from Sept. 1, 1639. From the fact that the name of Thomas Savage often occurs as a witness to instruments drawn by Lechford, I conjecture that he was a near neighbor, or perhaps a fellow-tenant under the same roof. Occasional entries like the following give glimpses of the interior of "the chamber etc.," and of Lechford's manner of living:—

1639. June. "Borrowed of Mr. Story about a month since 2^{li} & halfe of the beft fuger at 2^{sh} the pound   5$^{s.}$ 5$^{d.}$

April. "Rec^d of Mr. Keayne for a filver laced coate and a gold wrought cap   £2. 10.

May. "Received of Mr. George Story 4 yards and halfe a quarter of tuft holland to make my wife a waftcoate at 2^{sh} 8^d per yarde   11$^{s.}$

1640. Jan. 31. "I payd Nathaniel Heaton for full of writings & cutting wood   5$^{s.}$

Feb. 1. "I payd John Hurd, delivered to his wife by Sara our mayd, for making my wife's gowne   8$^{s.}$

"I payd Thomas Marfhall before hand for wood, delivered by my wife to his wife in the 10 moneth laft paft [Dec. 1639]   £1.
Since which time I had of him 6. loads of wood at 5$^{s.}$ fo I owe him   10$^{s.}$

Jan. 12. "Received of Mr. Keayne 6$^{li.}$ of Spanifh tobacco upon account. And I owe him 1 load of wood, a good load.

"I payd Mr. Burton for malt, cheefe, and irons, £1.—and owe him 8$^{s.}$ 9$^{d.}$—in 10th [month] last.

1641. "Mary Sherman came to my wife the twelveth day of Aprill, 1641."

Almost from the hour of his landing at Boston, he was regarded with distrust by those whose influence prevailed in state and church. First, because of his *profession;* for, to "some of the magistrates," and doubtless to Governor Winthrop himself, the employment of "lawyers to direct men in their causes," seemed more objectionable than the custom of obtaining advice from the judges on an *ex parte* statement before the public hearing of the cause.* Winthrop himself, Bellingham, Humphrey, Dudley, Downing,— and perhaps Pelham and Bradstreet,— had been students of law in England; but, on this side of the Atlantic, their legal knowledge was not called into requisition, except as it contributed to qualify them for seats in the Court of Magistrates or as legislators for the new colony; "no advocate being allowed," † and the exercise of the profession of an attorney being discountenanced so far as possible without absolute interdiction.

But Lechford was not only *professionally* but *doctrinally* objectionable. Though he came to New England, as he says, with a disposition to "lay aside all by-respects, to join with the Church here," "he could not be satisfied in diverse particulars," and "desired to open his mind in some material things of weight concerning the Christian faith" wherein he differed from the received belief of the Massachusetts churches. He was not long in giving to these points of difference more than a sufficient prominence. On his passage hither, he had discussed them with his fellow-passengers; and before, or soon after, his arrival, he made a written statement of his opinions and the arguments by which he sustained them, and placed the paper in the hands of Mr. Downing. ‡ These opinions, which he tells us he "did not lightly or hastily take up, but upon good grounds and mature deliberation, long before he ventured to betake himself into these parts of the world," § involved what magistrates and elders held to be fundamental errors, and such as prevented his reception to

---

* *Winthrop*, ii. 36. "No judge can be wise enough to decide always with satisfaction to both parties," observes Mr. Savage, "after privately hearing, and of necessity, as it were, undertaking the cause of one, before issuing process."
† *Ibid.*
‡ Letter to Edmund Browne, Dec. 10, 1638. *Journal*, p. 28.
§ To Hugh Peters, Jan., 1639. *Ibid.*, 30.

church fellowship. These errors, as stated by Mr. Cotton, were: " 1. That the Antichrist described in the Revelation was not yet come, nor any part of that Prophecy yet fulfilled from the 4th chapter to the end. 2. That the Apostolick function was not yet ceased: but that there still ought to be such, who should by their transcendent Authority govern all churches." *

Lechford himself conceived that his opinions on these controverted points "might be held, or not held, *salva fide*," † and without impediment to church fellowship with those of opposite belief. Indeed, modern orthodoxy, even of the most rigid type, would hardly insist on the identification of the pope of Rome with the prophetical antichrist, and a denial of the permanency of the apostolic function, as essential pre-requisites to church communion, or for the elective franchise. But to the elders of the Bay, in 1638,— when the churches had not yet escaped the dangers of Antinomianism nor been thoroughly purged of all the eighty-two errors condemned by the synod of the year previous,— every deviation from the established creed was matter of grave importance. Moreover, although Lechford professed a disinclination to controversy, he certainly took no great pains to avoid it; so that before he had been many weeks in the colony, his peculiar views were somewhat widely made known, not only through oral discussions, but by means of two or three manuscript volumes of his composition, which he had tendered for the perusal of some of the jealous guardians of orthodoxy in the churches.

In the letter to Hugh Peters, before quoted, Lechford writes: "I showed you my books: you had not leisure to peruse them. I likewise, long before, showed my main book to Mr. Cotton. He had not leisure to read it; and the first draught of that *Of Prophesie*, it lay in his house at least five weeks." Peters had too much work before him, in New England and Old, for wasting his time over the crude speculations of an honest but narrow-minded enthusiast; and Mr. Cotton was perhaps less zealous in

---

* *Way of Congr. Churches cleared*, pt. i. p. 71.

† See after, his Propositions to the General Court, June 11, 1639.

heresy-hunting, if not more tolerant of error, than before his own narrow escape from the censure of the synod of 1637 for an imimputed taint of Hutchinsonianism. If Lechford had gone no farther to look for readers and provoke criticism, he might have fared better, — might have found a way at last to the fellowship of the churches and the favor of magistrates, and have lived and died in Massachusetts, in comfortable circumstances and with a more favorable opinion of "rigid separations" and "electorie ways" than he has expressed in *Plain Dealing*. But, in an evil hour, he sought counsel of the deputy-governor, Thomas Dudley, a man whose conscientiousness was as morbid, his vision as narrow, and his prejudices as strong, as Lechford's own; who was so jealous for the purity of the faith that he magnified to a mountain every mole-hill of error, and saw in the toleration of new opinions a "cockatrice's egg,"

"To poison all with heresy and vice."

"After the court here ended," wrote Lechford to Hugh Peters, in January, 1639, "I delivered [my book] *Of Prophesie* to Mr. Deputy, to advise thereof as a private friend, as a godly man and a member of the Church, whether it were fit to be published. The next news I had was, that at first dash he accused me of heresy, and wrote to Mr. Governor that my book was fitter to be burned. . . ."

The court to which Lechford refers was probably the Quarter Court held at Boston, Dec. 4th, 1638. On the eleventh of the same month, Dudley wrote from Roxbury, to Winthrop: —

"Sir. Since my cominge home, I have read over Mr. Lechford's booke, and finde the scope thereof to be erroneous and dangerous, if not hereticall, according to my conception — His tenet beinge that the office of apoſtleſhip doth ſtill continew and ought ſoe to doe till Criſt's coming, and that a Church hath now power to make apoſtles as our Saviour Criſt had when hee was heere. Other things there are, but I pray you conſider of this, and the inſeparable conſequences of it: I heare that Mr. Cotton and Mr. Rogers know ſomethinge of the matter, or man, with whome you may if you pleafe conferre: I heare alfo that hee favoureth Mr. Lentall* and hath fo expreft himfelfe fince Mr. Lentall was queftyoned by the minifters: It is eafyer ftoppinge a breach when it begins, then afterwards: wee fawe our error in fufferinge Mrs. Huchinfon too longe. I have fent you the booke herewith that in ftead of puttinge it to the preffe as hee defireth it may rather be putt into the fire as I defire: But I pray you lett him know that I have fent the booke to you, that after you have read it (which I think you faid you had not yet done) it may be reftored to him.

. . . .

"I fuppofe the booke to be rather coppyed out then contryved by Mr. Lechford, hee beinge I thinck, not foe good a grecyan and hebritian as the author undertakes to be." †

Either Winthrop's zeal was less lively, or he saw less danger in the new heresy and its "inseparable consequences," than his colleague. Before the end of the month, Dudley wrote again: —

"For Mr. Lechford and his booke, you fay nothing, and I have fince heard that the worft opynion in his book (which I thinck I fhall proove to be herefy) is taken upp by others. Nowe feeing that this is the way Sathan invades us by (viz. new opynions and herefyes) it behooves us to be the more vigilant, and to ftirr upp our zeale and ftopp breaches at the beginninge, leaft forbearance hurt us as it did before." ‡

Lechford's character appears in a very favorable light in his comment on the course pursued by Mr. Dudley. After disavowing the chief heresy imputed to him, "though indeed my words might have been so strained," he adds: —

"I fpeak according to my light, and dare do no otherwife. If hotly [pressed by?] Mr. Deputy, I impute it to his zeal againft errors: I am not angry with him for it. But when I faw feven fhepherds and eight

---

* See *Plain Dealing*, pp. 22, 41, and notes 78, 144. Mr. Lenthall was "queftioned by the ministers," Dec. 11, 1638, at a conference (held at the house of Capt. Israel Stoughton, in Dorchester), of which some manuscript notes, taken by Robert Keayne, have been preserved.

† *Proceed. Mass. Hist. Soc.*, 1855-8, pp. 311, 312.

‡ Dudley to Winthrop, Dec. 29, 1638, in 4 *Mass. Hist. Coll.*, vii. 111.

principal men called out againſt me, as if I were an Aſſyrian [the allusion is to Micah, v. 5], I thought there might be ſomething in me to be reproved, and that it concerned me to look about me. I dealt plainly. ... Thereupon my book was referred to the conſideration of the Elders."

This reference to the elders was the occasion of his addressing to Hugh Peters, Jan. 3, 1638-9, the letter from which several extracts have already been introduced. In an interview with some of the magistrates, he had "intimated a word of [his] *other*, main book," treating of Antichrist and of the millenial kingdom of Christ. "They all now press me to produce *that*. I told them it was not ready for their view: I must fair write it, and alter some things: yet at length, upon promise that I should have it again (for if it be no error, I will not part with it for £100) I promised to let them see it. I have accordingly left it to Mr. Deputy and the Governor (who also desired to see it)." This book, with the one *Of Prophesie*, was to be submitted to an assembly of the Elders; and Lechford writes to request Mr. Peters that he would himself be one of the council, "Mr. Ward another, and Mr. Parker of Newbury; and that Mr. Norton and Mr. Phillips may likewise be called;" who should "soundly and maturely advise and consult of the matter," with "all lawful favour" to the writer.

I find no subsequent mention of this council, unless it be referred to by Mr. Cotton, in the passage already cited (from the *Way of the Congregational Churches cleared*, pt. i. p. 71), where Lechford is said to have been "dealt withall both *in conference* and (according to his desire) in writing." Neither mode of dealing was effectual to convince him of error, nor would the elders admit that his opinions might be held "*salva fide*." So he was compelled to remain without the church; and exclusion from church fellowship carried with it exclusion from the privileges of a freeman, and disqualification for civil office.

His professional ability was not inconsiderable; but the field for its exercise was restricted. "Kept from all place of preferment in the Commonwealth," he was "forced to get his living by writing petty things, which scarce found him bread," as he complained to his friends in England, after two years' residence here.* Though his imputed heterodoxy did not prevent his occasional employment, by those of sounder faith, as a conveyancer, scrivener, or draughtsman, his receipts for such professional services were pitifully small. His Journal contains not only the record of every instrument drawn by him while he was in this country, but an account of the compensation he received; from which it appears that his professional income, for the two years after his arrival, was a little more than £47; about £9 of which was in debts remaining unpaid in July, 1640.†

In June, 1639, when he had been nearly a year in Boston, he presented to the General Court certain propositions‡ for the regulation of civil actions, and for the recording of judicial proceedings. He had perhaps been encouraged to hope — for he states that his propositions were "made upon request" — that the Court, notwithstanding his incligibility to public office, would employ his services in the humbler capacity of clerk or public notary, and provide for his support by giving him work to do for which his studies and experience peculiarly qualified him. His application was not successful. "The Court was willing to bestow employment upon me," he writes (in short-hand) in his Journal, "but they said to me that *they could not do it for fear of offending the churches, because of my opinions*. Whereupon I thought good to propose unto them as followeth:" —

"Certaine Propoſicons to the generall Coʳt, 11. 4. 1639.
"Whereas I have delivered that Prophefying in the Church is properly, and therefore ought to be mainely, of propheticall fcriptures: and that Apoſtles, Evangeliſts, and Prophets ought to be continued as well and as long as Paſtors and Teachers or any other the undoubted officers, (by vertue of the Inſtitution, *ſome Apoſtles, ſome Prophets*, etc.) and that

---

* *Plain Dealing*, 69.
† "Money received upon my book, as appeareth, £38. 8. 5, or thereabout, befide in debts owing, £8. 18. 10. Caſt, 2 (5) 1640." — Short-hand note in *Journal*.
‡ Printed in *Plain Dealing*, pp. 29, 30.

it is probable there shall come yet a greater Antichrist then ever hath bin, etc.

1. I doe not refuse Church Communion wth any that hold the contrary.
2. If the Elders upon perufall of my books, and hearing me, will give their cenfure and reafons in writing or otherwife againft the maigne propofitions in my bookes, if they cannot fatisfy me fo farre as to recant, yet I fhall be content to be filent.
3. If the Elders upon perufall of my bookes, and hearing me, can convince me of error, in the maigne propofitions, I fhall be ready to retract, yea, to burne my bookes.
4. If the State and the Elders thinke that the matters I treate on are not *tanti*, or that they are iuft occafion of difturbance, I fhall be content they will advife of them 12. moneths or more, wth filence on my parte during that fpace, faving to the Elders and chiefe men, provided that I may have imployment to fubfift among you, and in the meane while be admitted to the privileges of God's houfe; for that all I write may be held, or not held, *falva fide*, as I conceive. Wth all due fubmiffion to this hono^{ble} Co^{rt}

Yo^r humble fuppliant
Tho: Lechford.

It was in response to this application, probably, that he was "dealt withal, *according to his desire*, in writing," as Mr. Cotton has mentioned. Whether or not the Court gave favorable consideration to the proposition by which Lechford engaged himself to refrain from controversy for twelve months, on consideration of receiving employment, does not appear. But whatever good intentions in his behalf the magistrates, or some of them, may have had, were counteracted by his own imprudence.

In the summer of 1639, he was employed by William Cole*
and his wife Elizabeth, for the prosecution of an action against
her brother, Francis Doughty, of Taunton, whom she charged
with having defrauded her of her marriage-portion and her share
in their father's estate. To the preparation of this case, Lechford's Journal and memoranda show that he gave much attention.
On the trial before a jury, at the quarter court in September, his
zeal for his clients betrayed him into an indiscretion (to use no
harsher term) which subjected him to the deserved censure of
the court, and gave occasion, not wholly displeasing to the magistrates perhaps, to prohibit him from the exercise of the profession of an *advocate*, to which, as has already been intimated, he
does not appear to have had any legitimate title. The order of
the court is in these words : —

"Mr. Thomas Lechford, for going to the Jewry & pleading wth them
out of Court, is debarred from pleading any man's caufe hereafter, unleſſe his owne, and admoniſhed not to p^{r}fume to meddle beyond what
hee ſhalbee called to by the Courte." †

Lechford submitted, in a good spirit, to this censure. A few
days after receiving it, he presented to the General Court a
petition for pardon, with a frank confession of his fault. Of this
petition he has preserved a copy, in short-hand.‡ It is worth
insertion here, as characteristic of the man.

---

* William Cole, who came from Chew-Magna, co. Somerset, married Elizabeth, daughter of Francis Doughty, a merchant and sometime alderman of the city of Bristol. Mr. Doughty died before 1637, and while William Cole and his wife were yet in England. Mention of his son, the Rev. Francis Doughty, is made in *Plain Dealing*, p. 41 (of this edition, p. 91, and note 136). John, a brother of William Cole, was living in Farrington, co. Somerset, in July, 1639. The names of William, John, and Nicholas Cole, appear among the early inhabitants of Mr. Wheelwright's plantation at Exeter, and that of William is subscribed to the association of Exeter planters, Oct. 4, 1639 (Hazard, i. 463).

† *Mass. Col. Records*, i. 270.

‡ *Journal*, page 117.

INTRODUCTION.

"*To the Honble the Governor, Council and Affiftants of this Jurifdiction and to the General Court thereof affembled*, 10. 7. 1639.

"The humble fupplication or petition of Thomas Lechford, [late of Clement's Inn in the County of Middlefex, gent.]*

"Truely fhowing and aknowledging that he did offend in fpeaking to the Jury without leave, in the caufe of William Cole and his wife; and fo much the more inexcufable was this delinquency inafmuch as he knew it was not to be done by the law of England. Yet he conceiveth it was not Embracery, for that he had no reward fo to doe; and fome extenuation may, he conceiveth, be gathered by one or two feeming approbations of the like which he hath obferved in other caufes here. Notwithftanding, he is heartily forry for his offence, and acknowledgeth the juftice of this Court, and is comforted in this — that he hopeth it may doe him good and the example be a benefit to the publick. Touching his fpeaking in publick for future time, he fubmitteth to the wifdom of the Court; and for that which is paft, he came to the Court being retained, and it's true ftood there at the lower end, next the deputy Marfhal, attending unto a caufe or two wherein your petitioner was retained. It was to fhow his readinefs to do the countrey any fervice he might, as well as to get a little money for himfelf. Some fpeeches of his, fpecially fome involuntary and of fudden *interruptions* of fome in *authoritie*† being made, whereof fome might be occafioned by himfelf, [being too *tartly*, as he conceiveth, rebuked and *hindered* by fome of the Court,]‡ and zeal of fpeaking for his mafters, may feem to offend fuch as have not been accuftomed to publique pleadings of advocates. Such *expreffions* of his and involuntary offences he humbly prayeth may be paffed by; and fuch occafions of pleadings your fuppliant will readily forbear, as not being fufficient or inclinable by nature thereunto. And he hopeth that this Court and country may upon trial of this petitioner in fome other *ufe* find him, as in many things *ignorant*, fo teachable and tractable.

"In the mean while, if your petitioner hath any the leaft talent to doe

---

* The words included between brackets were croffed out on revifion.

† The characters are so closely crowded together, and rendered so indistinct by the spreading of the ink on the thin paper, that a few words are quite illegible, and of two or three others the reading is doubtful. For the former, I have left a blank space; and the latter are printed in *italics*.

‡ Several words were crossed out here, others interlined, and these in turn crossed out; and the sentence appears to have been finally left incomplete.

INTRODUCTION.

you any fervice in a way of profitting himfelf [ ] livelihood, he defireth it. He is heartily ready, and humbly prayeth the fame, in regard of his low and poor eftate, not unknown to fome of your Worfhips : Unfeignedly defiring both to live and die with you in the way of God's ordinances, wherein your petitioner hopeth in fome good time or other fome of the reverend Elders and himfelf may come to a perfect or at leaft a fair underftanding of each other, which that we may do is the unfeigned daily prayer of your unworthy petitioner,

"Thomas Lechford."

His submission was probably accepted by the Court, and he was suffered to return to the practice of his profession as an attorney, which, under the restrictions imposed upon it, promised little improvement of his "low and poor estate."

In the autumn and winter of 1639, he received some slight assistance, in the way of employment, from the magistrates. For Mr. Endicott, he had written "The Court booke,* at 16ᵈ· a sheete, 102 sheetes," and received £6. 16s. some time in June or July. In November, after the surrender to Massachusetts of the Dover patent, he wrote "For the Country: The writing of receipt of the Inhabitants of Dover and Kittery and Oyster River into the Protection of this Jurisdiction: The Commission to Mr. Bradstreete for those places: The institution and limitation of the Councell of this Jurisdiction: Another of the same: *Charta libertatis*: The Act of the publique and private tenure of land: The division of the Plantation into shires:" for all which he received the sum of *eleven shillings*.† Not long afterwards, he was employed in the more important task of transcribing the

---

* I cannot learn that this copy of the "Court Book" has been preserved. It was, undoubtedly, a transcript of the Colony Records, made for Mr. Endicott's own use or for that of the Salem Quarter Court. A. C. Goodell, Esq., of Salem, to whom I applied in the hope of discovering some trace of this volume, calls my attention to the agreement of the number of "sheets" with the folios of the Colony Records, from the first court at Charlestown, Aug. 23, 1630, to the end of the Quarter Court at Boston, June 4, 1639, making 202 pages (55-256 of the first volume of the manuscript Records of the Governor and Company; pp. 73-268 of the *printed* Records), or 101 folios.

† *Journal*, p. 139.

"breviat of laws," subsequently adopted, with some amendments, as the Body of Liberties.* While engaged in this work,— which, in his hands, we may be sure was something more than that of mere transcription, — he could not resist the temptation, or, as he chose to express it, "he conceived it his duty, in discharge of his conscience," and "as *Amicus curiæ*, with all faithfulness to present" to the Governor and magistrates his objections to certain laws proposed to be embodied in the code.

In May, 1640, in "a paper *intended* for the honored John Winthrop," he expressed his convictions of the advantages and the necessity of submission to the King, and acknowledgment of the authority of the Church of England, "if it be but by way of advice;" frankly confessing that for himself he "disclaimed Parker" and "inclined to Hooker and Jewel as to government." † After this paper was drawn, Dudley was elected governor; and it is not likely that Lechford transferred to him the good advice prepared for Gov. Winthrop.

The year during which he had conditionally promised to keep silence, "saving to the Elders," on matters of difference between himself and the churches, had now expired. He had been "seriously dealt withal," and had been indulged in his desire for "reasons in writing." ‡ But his hope that "in some good time the reverend Elders and himself might come to a perfect, or at least a fair understanding," was less and less likely to be realized. He was becoming more dissatisfied with the condition of affairs in New England, both in church and commonwealth. In July, 1640, he wrote to England: "I know my friends desire to know whether I am yet of any better mind than some of my actions about the time of my coming away did show me to be. I do profess that I am of this mind and judgment, I thank God: that Christians cannot live happily without Bishops, as in England, nor Englishmen without a King. Popular elections indan-

---

* See *Plain Dealing*, p. 27 (this edition, p. 64, and note 91) and p. 31 (this edition, 72, and note 101).

† *Plain Dealing*, pp. 34-37.

‡ See before, p. xxvi.; and *Plain Dealing*, p. 77.

ger people with war and a multitude of other inconveniences." * Of the people of Massachusetts he says, "I am not of them, in church or commonweal. Some bid me be gone: others labor with me to stay fearing my return will do their cause wrong; and loth am I to heare of a stay, but am plucking up stakes with as much speed as I may, if so be I may be so happy as to arrive in Ireland, there at least to follow my old profession," &c. " Some silence my letters and will not dispute with me, I think either out of distrust of me, or else despaire of their cause ; some cry out of nothing but Antichrist and the Man of Sin. . . . But few know my full mind in some things of weight whereof I do professe I was ignorant and misled in England. You may wonder how I am now reformed," &c.

"I never intended," he writes, "openly to oppose the godly here in any thing I thought they mistooke." † If he maintained some reserve in the expression of his "full mind in some things," he certainly made no secret of his dislike of "electory ways" and of congregationalism, as is evident from the advice which he proffered to the governor and magistrates, and from his queries propounded to the Elders of Boston, which challenged a discussion of the nature and constitution of a church and the validity of congregational ordination. ‡

That his opinions, and his zeal in advocating them, made him obnoxious to the magistrates, as well as to the Elders, is no matter of surprise. When the course which had been taken with others who had similarly offended is considered ; when it is remembered that, not only had teachers of doubtful orthodoxy, like Roger Williams and Wheelwright and Mrs. Hutchinson, been banished from the jurisdiction, but laymen of influence and position, like Stoughton and Aspinwall and Coggeshall, when suspected of a taint of heresy or " sedition," had been as summarily and as severely dealt with,— the leniency shown to

---

* Short-hand copy, in *Journal*, p. 159.
Comp. *Plain Dealing*, pp. 68, 69.

† See *Plain Dealing*, p. 77.
‡ *Ibid*, p. 55.

INTRODUCTION.

Lechford is remarkable. It could hardly have been from motives of policy — only his own vanity could have suggested that it was from "fear his return would do their cause wrong" — that he was suffered to remain so long unmolested. It must rather have been owing to a conviction of his honesty, his conscientiousness, and, possibly, to his lack of influence and the slight danger of infection by his teachings. It would not be easy to find, in the first fifty years of the history of Massachusetts, another instance of so great tolerance of opinions so radically opposed as were Lechford's to the views of the founders of the colony, and so subversive of the constitution of civil government and of the church polity they sought to establish in New England. He was neither a freeman nor a church-member; not even a householder; in the eye of the law he was merely a "transient person," who might be driven away with slight ceremony. His calling made him unwelcome; his creed, in the judgment of others besides Thomas Dudley,* was "erroneous and dangerous, if not heretical." He questioned the validity of any non-episcopal ordination, and saw, in the exercise by the people of the right to elect their own rulers, the root of all evil. He would not acknowledge "a church without a bishop," and did not hesitate to express his belief that all was going wrong, and must go worse, in "a state without a king." In the complacent consciousness of his own clearer light and well-grounded convictions, he felt it to be his duty to point out to Governor Winthrop, to Mr. Wilson, and to Mr. Cotton, the errors wherein through ignorance they had gone astray, and were misleading others.† That he should have been permitted for two years and a half to hold his course unchecked, and that his unconcealed and somewhat *aggressive* dissent should have so long escaped censure,

---

* See before, p. xxii.

† "O mercy, mercy, from all the powers of mercy in heaven and earth"— he wrote in 1640 — "to such as sin of ignorance!" And against this, he modestly noted in the margin : "In the number of the ignorant I hold *myself*, and Mr. Burton, Mr. Prynne and Dr. Bastwick, and a multitude more." *Journal*, p. 159 (in short-hand).

INTRODUCTION.

proves that the founders of Massachusetts were not incapable of the exercise of toleration, even though they might not give it a place among the virtues.

At length, however, their patience was exhausted. In September, 1640, for a new offence, with which his questioning of the Boston elders * may have had something to do, he was presented by the grand jury, and summoned before the Court of Magistrates in December. When the General Court was in session (Oct. 7), they were "pleased to say something to him, as for good counsel about some tenets and disputations which he had held; advising him to bear himself in silence and as became him." A few weeks afterwards, he writes in his Journal: "I am summoned to appear in court to-morrow, being the first of 10th, 1640. The Lord God direct me, &c." In a letter to England, dated Dec. 19, he mentions having been "lately taken at advantage and brought before the Magistrates, before whom, giving a quiet and peaceable answer [he] was dismissed with favour," &c.† Of this answer he preserved a copy, or perhaps the original draft, in short-hand, in his Journal. An extract from it is printed in a note on page 157 of this volume. Confessing that he had "too far meddled in some matters of church government and the like, which [he was] not sufficient to understand or declare," he threw himself on the mercy of the court. His submission was accepted, and the record shows that —

"Mr. Thomas Lechford, acknowledging hee had overſhot himſelfe, and is ſorry for it, promiſing to attend his calling, and not to meddle wᵗʰ controverſies, was diſmiſſed." — *Mass. Col. Records*, i. 310.

Mr. Savage, in a note to Winthrop (ii. 36), cites this as a "curiosity in legislative and judicial economy." He was under the

---

* See *Plain Dealing*, p. 55 (this edition, p. 128).

† "Our chiefe difference was about the foundation of the Church and Miniſtery, and what rigid ſeparations may tend unto, what is to be feared, in caſe the moſt of the people here ſhould remaine unbaptized; &c." *Pl. Deal.*, 77 (this ed. 156–7).

impression that the engagement "not to meddle with controversies" was inconsistent with the promise "to attend his calling," since "the very calling by which he sought to earn his bread was that of an attorney." The inconsistency disappears on learning from Lechford himself that he was brought before the quarter court on the presentment of a grand jury, and that the controversies in which he had "too far meddled" concerned "matters of church government and the like,"—"the foundation of the church and the ministry, and what rigid separations may tend unto." He acknowledged his fault, promised amendment, and the court dismissed the complaint. Lechford certainly did not feel that he had been hardly dealt by. He avers that he was "dismissed with favour, and respect promised him by some of the chiefe, for the future." *

Sometime in 1640, he was enrolled in the "Military Company of Massachusetts," afterwards the "Ancient and Honorable Artillery." He perhaps owed his election to his intimacy with Thomas Savage, one of the original members of this company, and to the friendship of the captain, Robert Keayne.

Among those with whom Lechford appears to have been on very friendly terms, was George Story, "a young merchant of London," as Winthrop calls him, who lodged in the house of Richard Sherman, and who was the chief instigator of the proceedings against Capt. Keayne in the famous "sow case." For six or seven years from its commencement in 1636, this "great business upon a very small occasion" divided the people of Boston into factions, disturbed the peace of the churches, had an influence in elections, awakened a "democratical spirit" throughout the colony, and at last (in 1643) came near bringing about a radical change in the constitution of the General Court, by depriving the magistrates of the exercise of a negative voice on the action of the house of deputies.† In 1641, the quarrel had not yet reached its height, but it had already assumed for-

---

* *Plain Dealing*, 77. † See *Winthrop*, ii. 69–71, 115–119.

midable proportions. That Lechford should become implicated in it, was inevitable. The only attorney in Boston, and the common friend of Story and of Keayne, he received the confidences of both parties, tried his hand at peace-making, gave advice to both, and, of course, offended both; besides exposing himself to the suspicion of wrong-dealing. The trouble which this affair occasioned him may have contributed to hasten his return to England. About a week before he sailed from Boston, he drew up a statement of his connection with the case, for the purpose of clearing himself of "divers imputations" of having promoted litigation by advice which, "in the simplicity of his heart," he had given to Mr. Story and Goody Sherman. This paper is dated July 24, 1641. In the first draft (in his Journal), he had written: "Being purposed some time at least to visit my native" — ; but drew his pen through the unfinished sentence, and interlined, in its place: "*Now* being purposed, God willing, to visit my friends in England." In another paragraph, alluding to a conversation which he had with Story, "one Lord's day when the Sacrament was at Boston," he fixes the time by adding, "being the next day as I remember after the newes *that it was supposed Mr. Prynne had sent me money for my passage.*"

Mr. Cotton says that Lechford, "when he saw he could not defend the Error [that the Apostolick function was not yet ceased] but by building again the Bishops, against whom he had witnessed (as he said) in soliciting the cause of Mr. Prynne, he rather then he would revoke his present tenent, acknowledged he was then in an Error when he took part with Mr. Prynne and Mr. Burton, and *therefore he would now return to England* again, to reduce those famous witnesses from the Error of their way. And accordingly, away he went." *

On the same day on which he wrote the statement above-mentioned, Lechford made a letter of attorney to Thomas Savage, to receive all moneys due him in New England, and all letters which should be sent to him, "and the same letters to peruse,

* *Way of Congr. Churches cleared*, pt. i. p. 71.

and send and return them and the said moneys and debts to him, in money or goods and commodities," &c.*

The last entry in his Journal, before leaving Boston, was made on or after July 29. It is a memorandum of his obligation by bond (in which Mr. David Offley was his surety) to Mr. Joshua Hewes of Roxbury, to pay £8. to "Mr. Joshua Foote at the Cocke in Grace church Streete," before Christmas, on a bill or note dated July 27.

On the opposite page are two unimportant entries, of payments of money in England, in the discharge of commissions intrusted to him before sailing. At the head of this page is the date, "Post Mich[aelmas], 17 Car. 1641."

The vessel in which he took passage from Boston sailed on the third of August. We learn from Winthrop (ii. 31), that among her forty passengers were John Winthrop, Jr., Hugh Peters, Thomas Welde, and William Hibbins, who, "finding no ship which was to return right for England, they went to Newfoundland, intending to get a passage from thence in the fishing fleet. ... They arrived there in 14 days, but could not go altogether, so were forced to divide themselves, and go from several parts of the island, as they could get shipping." Lechford mentions having "touched, coming homeward," at Newfoundland.† On the 16th of November, he was once more an inmate of Clement's Inn, and had "returned humbly to the Church of England." ‡

From this time, his personal history remains unknown. The address "To the Reader" of his book, dated Jan. 17, 1641-2, is the last trace of the author which he has left us. All that we have to add is comprised in a single sentence by Mr. Cotton: —

"When he came to England, the Bifhops were falling, fo that he loft his friends, and hopes, both in Old England and New: yet put out his Book (fuch as it is) and foon after dyed."— *Way cleared*, pt. i. p. 71.

* *Journal*, p. 234.   † *Plain Dealing*, p. 46 (this edition, 109).   ‡ *Ibid.*, 68.

That the magistrates and ministers of Massachusetts should not look with favorable regard upon the book or its author, was natural ; but it is not easy to discover good grounds for so severe a judgment as that recorded by Mr. Cotton upon "Plaine dealing, which (in respect of many passages in it) might rather be called false and fraudulent." Lechford was not a man of broad views, or of great political sagacity. He was tolerably clearsighted, but not far-sighted ; a good observer, but a bad prophet. His own reverses had apparently taken from him whatever hopefulness he had by nature, and he looked habitually to the darker side. Such men cannot lead colonies, or found States. He was out of place in New England, and would have been none the less so, if he had been as firmly convinced as was Mr. Cotton of the identity of the Church of Rome with Antichrist. Little as Winthrop or Cotton could foresee of the future of New England,—of the ultimate results of the work in which they were engaged,—Lechford foresaw less. To his view, prejudiced somewhat, no doubt, by the adverse circumstances against which he struggled from first to last in Massachusetts, "all was out of joint both in Church and Commonwealth ;" * nothing better was to be anticipated from popular government than anarchy and bloodshed ; from separatism, than a speedy relapse to heathenism ; and from a disregard of "worthy *lawyers* of either gown," than tardy repentance.† There were, he thought, "*some* wise men" in New England ; but "wiser men than they," if they had attempted to set up in a wilderness a " strange government, differing from the settled government [in England], might have fallen into greater errors." The only hope he saw for the country was in the exertion of the king's prerogative, and the extension of the authority of the Church. "With some kind of subjection or acknowledgment of authority of the Ministry in England," then perhaps, "under God and the King," the colony might "make Church-work and Common-wealth work indeed, and examples to all Countries." ‡

* *Plain Dealing*, p. 71.   † *Ibid.*, 28.   ‡ *Ibid.*, pp. 34, 35.

Yet *Plain Dealing* was not written in an unfriendly spirit. "I doe not this, God knoweth," says the author, "as delighting to lay open the infirmities of these well-affected men, many of them my friends,—but that it is necessary, at this time,"—when England was in danger of falling into the same kingless and churchless abomination of desolation,—"for the whole church of God, and themselves, as I take it."* However prejudiced in his judgments, however unwarranted his inferences, in his record of *facts* he is conscientious, painstaking, tolerably exact, and almost always reliable. And this it is which gives to his book its peculiar value. It is a view of New England,—more particularly of Massachusetts,—taken upon the spot by an intelligent observer, who, though unsympathising, was not in the main unfriendly; and who, while he certainly did "naught extenuate," cannot justly be charged with setting down aught in malice. His mistakes are comparatively unimportant; and the information he gives of the state of the country, civil and religious, from 1638 to 1641, is valuable enough to render his book nearly indispensable to the study of New-England institutions.

The Massachusetts Historical Society possesses a manuscript copy of a part of *Plain Dealing*, of which the Hon. James Bowdoin, in a note to the Society's reprint of the volume, gave the following description:—

"The MS. was at some former period bound up with others, and was probably at that time perfect. It now consists but of twenty-nine pages in small 4to. It is obviously ancient, whether we examine the appearance of the paper, of which the water-marks cannot be distinctly ascertained, or the color of the ink, or the character of the hand-writing; which last is remarkably fine of its kind. The *shorthand*, of which there are short passages on pages 9, 16, 23, 24, and 27 [corresponding with pages 12, 20, 37–38, 39, 41, of the first edition of *Plain Dealing*], differs from any one that the writer has been able to find; and he re-

---

* *Plain Dealing.* "To the Reader;" (this edition) p. 7.

grets to add, that application to two members of our Society, who are accustomed to shorthand of many periods, has ended, like his own exertions, in an inability to furnish a translation of them. . . . That the MS. was written prior to the printed copy, seems certain, as well from these last considerations, as from the additions and verbal differences that distinguish the two copies:—That it was written *after* Lechford returned to England, is ascertained by its containing the passage, on p. 73 [first edition, p. 13], alluding to his having left New England the August preceding. . . .

"The MS. begins with its own page 7, which is page 8 of the Ebeling copy [of the first edition], at the words—'the Elders formerly mentioned. Then the Elder requireth,' &c. It ends with its own page 36, being [page 53, line 10, of the first edition], with the word 'perfected.'"*

Mr. Bowdoin gave reasons for concluding that this MS. "could not have been the *identical original* which Lechford eventually enlarged, nor that from which the printer copied;" and that "it was probably a duplicate original, made and deposited for security, lest the fruit of his labor should be lost, by fire or other accident." The handwriting of the MS. is unmistakably Lechford's, as a comparison of it with his Journal shows. It certainly was not the first draft or sketch of his book: the penmanship is too neat, and there are too few of the interlineations or erasures which abound on the pages of his Journal. My impression is, that the copy of which this is a part was one intended for the use of the printer; but that, on his passage homeward or after his return, the author found so much to amend and so much new matter to add, that it became necessary to make *another* revised copy, from which the book was printed. The additions and alterations, amounting (as Mr. Bowdoin states) to near one-half of the whole, were made, in some places, in *short-hand*, on the margins or blank spaces of the manuscript, and afterwards incorporated in the text,† or printed as notes.

In the note referred to, Mr. Bowdoin has given the results of

---

* 3 *Mass. Hist. Coll.*, iii. 397, 398, 400.   † See after, p. 57, note 77.

INTRODUCTION.

a careful collation of the Society's MS. with the printed volume. Of this collation I have made free use in the notes to the present edition; and, relying upon its accuracy, I have cited the manuscript as "Mass. Hist. Society's Manuscript," or "M.H.S. MS."

The system of short-hand which Lechford used was substantially that taught by John Willis, first published in 1602, and very popular in England for thirty or forty years afterwards. When the characters are well formed, not too much crowded, nor too minute, there is no great difficulty in deciphering them. Lechford was so familiar with this system, and so practised in its use, that he was not very careful how he wrote it, especially in his first drafts; and when, as on some pages of his Journal, he used bad ink on imperfectly-sized paper, it is not easy always to distinguish his circles from ellipses, straight lines from curves, or dots from dashes.

A second edition of *Plain Dealing* — or a re-issue of the edition of 1642, with a new title-page — appeared in 1644, as *New England's Advice to Old England*. I have never met with this edition, and mention it here only on the authority of Watt and Lowndes.

The copy which I have used while preparing this edition, and for the correction of the press, is from the library of George Brinley, Esq., of Hartford, to whom I am also indebted for the opportunity of consulting several rare tracts cited in the notes.

J. H. T.

HARTFORD, JAN. 8, 1867.

PLAIN DEALING:
OR,
NEVVES
FROM
New-England.

(Vivat Rex Angliæ Carolus,
Vivat Anglia,
Vivantq̃ eorum Amici omnes.)

A ſhort view of NEW-ENGLANDS preſent Government, both Eccleſiaſticall and Civil, compared with the anciently-received and eſtabliſhed Government of ENGLAND, in ſome materiall points; fit for the graveſt conſideration in theſe times.

By THOMAS LECHFORD of *Clements Inne*, in the County of *Middleſex*, Gent.

Levis eſt dolor, qui capere conſilium poteſt,
Et elepere ſeſe; Magna non latitant mala. Sen.

L O N D O N,
Printed by *W. E.* and *I. G.* for *Nath: Butter*, at the ſigne of the pyde Bull neere S. *Auſtins* gate. 1642.

TO THE READER.

*Very man is to approve himselfe, and anfwer to God for his actions his confcience leads him to; and next, to good men, as much as in him lyeth.* I have thus prefumed to enter into publique, for thefe reafons:

Firft, becaufe it is well knowne unto many, that heretofore I fuffered imprifonment, and a kind of banifhment out of this good Land, for fome acts conftrued to oppofe, and as tending to fubvert Epifcopacie, and the fetled Ecclefiafticall government of England: therefore now I defire to purge my felf of fo great a fcandall; and wherein I have offended, to intreat all my Superiours, and others, to impute it rather to my ignorance, for the time, then any wilfull ftubbornneffe.

Secondly, feeing that fince my comming home, I find that multitudes are corrupted with an opinion of the unlawfulneffe of the Church-government by Diocefan Bifhops, which opinion I beleeve is the root of much mifchiefe; having now had experience of divers governments, I fee not how I could with faithfulnesse to God, my King and Countrey, be any longer silent, especially considering some of these late troubles occasioned, among other sins, I fear, much through this evill opinion. Happy | shall I be, if any be made wiser by my harmes; I wish all men to take heed, how they shake hands with the Church of God, upon any such heedlesse grounds as I almost had done.

Thirdly, that I might (though unworthy) in a fit season, acquaint the learned and pious Divines of England with these my slender observations, quæres, and experiments, to the end they may come the better prepared, upon any publique occasion, for the consideration of such matters, and so at length, those good things that are shaken among us may be established, and truth confirmed.

It is enough for me, being a Student or Practiser at Law, faithfully to put a Case, which will be this: Whether the Episcopall Government by Provinciall and Diocesan Bishops, in number about 26. in England, being, if not of absolute Divine authority, yet nearest, and most like thereunto, and most anciently here embraced, is still safest to be continued?

Or a Presbyterian government, being (as is humbly conceived) but of humane authority, bringing in a numerous company of above 40000. Presbyters to have chiefe rule in the keyes, in England, be fit to be newly set up here, a thing whereof we have had no experience, and which moderate

To the Reader.

wife men think to be leſſe conſonant to the Divine patterne, and may prove more intolerable then the ſaid Epiſcopacie? Or an independent government of every congregationall Church ruling it ſelfe, which introduceth not onely one abſolute Biſhop in every Pariſh, but in effect ſo many men, ſo many Biſhops, according to New-Englands *rule, which in* England *would be Anarchie & confuſion?*

I would entreat thoſe that ſtand for this laſt mentioned manner of government, to be pleaſed to conſider,

1. *That the very terme of* leading, *or* ruling *in the Church, attributed to* Elders, *forbids it; for if all are Rulers, who ſhall be ruled?*

2. *The maine acts of Rule conſiſt of receiving into the Church by Baptiſme, or otherwiſe, and ejection out of the Church by cenſure, binding and looſing; now theſe are committed to the Apoſtles, and their ſucceſſors, and not to all the members of the Church.*

3. *All have not power to baptize, therefore not to receive into the Church, nor to caſt out of the Church.* My brethren, be not many maſters, *ſaith S. Iames,* 3. 1. The words of the wiſe are as goads, and as nayles, faſtened by the maſters of aſſemblies, which are given from one Shepheard, *Eccleſ.* 12. 11.

And whereas ſome may ſay, that this power of ruling is but miniſterially in the officers, and initiatively, concluſively, and virtually in the people: If ſo, what power ordi-

narily have the people to contradict the ministeriall works and acts of their Officers? Must the whole Church try all those whom their Ministers convert abroad, suppose among Indians, before they may baptize them? How can all the Church examine and try such? All have not power, warrant, leisure, pleasure, ability, for, and in such works, nor can all speake Indian language.

Doubtlesse the acts of rule by the Officers is the rule of the whole Church, and so to be taken ordinarily without contradiction, else there would be no end | of jangling: And thus taken, the whole Church of Corinth, *by S.* Pauls *command, (sc. by their Ministers) were to put away that wicked person, and deliver him up to Satan,* 1 Cor. 5. 13. *and restore him, and forgive him,* 2 Cor. 2. *and so all the doubt on that Text is (neer I think) resolved.*

Now that the government at New-England *seemeth to make so many Church-members so many Bishops, will be plaine by this ensuing Discourse: for you shall here find, that the Churches in the* Bay *governe each by all their members unanimously, or else by the major part, wherein every one hath equall vote and superspection with their Ministers: and that in their Covenant it is expressed to be the duty of all the members, to watch over one another. And in time their Churches will be more corrupted then now they are; they cannot (as there is reason to feare) avoid it possibly. How can any now deny this to be Anarchie and confusion?*

Nay, say some, we will keep out those that have not true grace. But how can they certainly discerne that true grace, and what measure God requireth? Besides, by this course, they will (it is to be feared) in stead of propagating the Gospel, spread heathenisme; in stead of gaining to the Church, lose from the Church: for when the major part are unbaptized, as in twenty years undoubtedly they will be, by such a course continued, what is like to become of it, but that either they may goe among their fellow-heathens the Indians, or rise up against the Church, and break forth into many grievous distempers among themselves? which God, and the King forbid, I pray.

And that you (courteous Reader) may perceive I have from time to time dealt cordially in these things, by declaring them impartially to my friends, as I received light, I shall adde in the last place certaine passages out Letters, sent by me into England *to that purpose, and conclude.*

And I doe not this, God knoweth, as delighting to lay open the infirmities of these well-affected men, many of them my friends, but that it is necessary, at this time, for the whole Church of God, and themselves, as I take it: Besides, many of the things are not infirmities, but such as I am bound to protest against; yet I acknowledge there are some wise men among them, who would help to mend things, if they were able, and I hope will do their endeavours. And I think that wiser men then they, going into a wildernesse to set up another strange government differing from the setled government here, might have falne into greater errors then they have done.

Neither have I the least aime to retard or hinder an happy and desired reformation of things amisse either in Church or Common-wealth, but daily and earnestly pray to God Almighty, the God of Wisdome and Counsell, that he please so to direct his Royall Majesty, and his wise and honourable Counsell, the high Court of Parliament, that they may fall upon so due and faire a moderation, as may be for the glory of God, and the peace and safety of his Royall Majesty, and all his Majesties dominions, and good Subjects. Vale.

Clements Inne,
    Jan. 17. 1641.

Thomas Lechford.

A TABLE of the chiefe Heads
of this DISCOURSE.

1. THe Church-government and administrations in the Bay of the Mattachusets. Page 2.
2. Their publique worship. 16
3. Touching the government of the Common-wealth there. 23
4. Certaine Propositions to the generall Court, concerning recording of Civill Causes. 29
5. A Paper of the Church her liberties. 31
6. A Paper intended for the Worshipfull John Winthrop, Esquire, late Governour, touching baptizing of those they terme without, and propagation of the Gospel to the Infidel Natives. 34
7. The Ministers and Magistrates their names. 37
8. The state of the Countrey in the Bay and thereabouts. 47
9. A relation concerning the Natives or Indians. 49
10. Some late occurrences touching Episcopacie. 53
11. Three Questions to the Elders of Boston, and their Answers. 55
12. A Paper of exceptions to their government. 56
13. Forty quæres about planting and governing of Churches, and other experiments. 58
14. An abstract of certaine Letters. 68
15. The Conclusion. 78

Plaine dealing:
OR,
NEWES
FROM
NEW-ENGLAND.

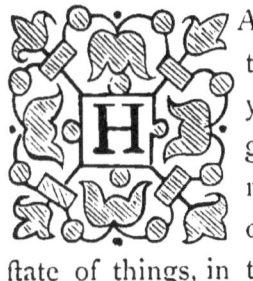Aving been forth of my native Countrey, almoſt for the ſpace of foure yeeres laſt paſt,[1] and now through the goodneſſe of Almighty God returned, many of my friends defiring to know of me the manner of governments, and ſtate of things, in the place from whence I came, *New England;* I thinke good to declare my knowledge in

[1] Lechford landed at Boſton, June 27, 1638. He ſailed for England, by the way of Newfoundland. Aug. 3, 1641. His "Quæres about Church government," in this volume, are dated from Clement's Inn, Nov. 16, 1641.—*Ms. Journal,* p. 1; *Plaine Dealing,* 13, 68; *Winthrop,* ii. 31.

such things, as briefly as I may. I conceive, and hope, it may be profitable in these times of disquisition.

2   *For the Church government, and administrations, in the Bay of the Mattachusets.*

<small>How Churches are gathered there.</small>

A Church is gathered there after this maner: A convenient, or competent number of Christians, allowed by the generall Court to plant together, at a day prefixed, come together, in publique manner, in some fit place, and there confesse their sins and professe their faith, one unto another, and being satisfied of one anothers faith & repentance, they solemnly enter into a Covenant with God, and one an other (which is called their Church Covenant, and held by them to constitute a Church) to this effect: *viz.*

<small>Their Church Covenant.</small>

To forsake the Devill, and all his workes, and the vanities of the sinfull world, and all their former lusts, and corruptions, they have lived and walked in, and to cleave unto, and obey the Lord Jesus Christ, as their onely King and Lawgiver, their onely Priest and Prophet, and to walke together with that Church, in the unity of the faith, and brotherly love, and to submit themselves one unto an other, in all the ordinances of Christ, to mutuall edification, and comfort, to watch over, and support one another.

<small>Election of their Church Officers.</small>

Whereby they are called the Church of such a place, which before they say were no Church, nor of any Church

except the invifible: After this, they doe at the fame
time, or fome other, all being together, elect their own
Officers, as Paftor, Teacher, Elders, Deacons, if they have
fit men enough to fupply thofe places; elfe, as many of
them as they can be provided of.

Then they fet another day for the ordination of their faid officers,[2] and appoint some of themfelves to impofe hands upon their officers, which is done in a publique day of fafting and prayer. Where there are Minifters, or Elders, before,[3] they impofe their hands upon the new Officers: but where there is none, there fome of their chiefeft men, two or three, of good report amongft them, though not of the Miniftery, doe, by appointment of the faid Church, lay hands upon them.[4] And after the faid

Their Ordinations.

---

[2] "Of this they give notice to all the near adjoining churches, ... intreating their prefence, and brotherly counfell, and affiftance. ... They give notice alfo thereof unto the governor, and fuch other of the magiftrates as are near to them, that the perfon to be chofen, meeting with no juft exception from any, may find the greater incouragement and acceptance from all." — Cotton's *Way of the Congreg. Churches*, 40. Comp. *Anfwer of the Boſton Elders to Lechford's Queſtions*, p. 55, *poſt*.

[3] "It is a received practife amongft us, that when any combine into a Church, there is one at leaft of them indued with able parts of humane and divine learning, that either hath been a Minifter in our native countrey, or is fit to be one amongft them, who ufually and frequently preacheth to them after they are united." — T. Welde, *Anfwer to W. R. his Narration*, (Lond. 4to. 1644,) p. 37. "It is our ufual and conftant courfe (as hath beene faid) not to gather any church, untill they have one amongft themfelves, fit for a Minifter, whom with all fpeede they call into Office, and account themfelves a lame and imperfect body till that be effected." — *Id.*, 55.

[4] "One of the Elders of the church,

Plaine dealing,

ordination, if there are any Elders of other Churches present, (as of late I have knowne divers have been present, under the names of the Messengers of the Churches) they give the new Officers the right hand of fellowship, taking them by the right hand, every one severally, or else, sometimes, one forraine Elder, in the name of all the rest, gives the right hand of fellowship, with a set speech unto them.⁵ Notice is given in divers Churches or other places, before-hand, of the gathering of every

<small>The right hand of fellowship by messengers of Churches.</small>

(if they have any) if not, one of the graver brethren of the church, (appointed by themselves to order the work of the day) . . . he then, with the Presbytery of that church (if they have any, if not, two or three others of the gravest Christians amongst the brethren of that church, being deputed by that body) do, in the name of the Lord Jesus, ordain him unto that office, with imposition of hands," &c. — Cotton's *Way of the Congreg. Churches*, 41. Comp. his *Keys of the Kingdom of Heaven*, &c., 21, 28, 37. See Mr. Gott's account of the ordination of Mr. Skelton, at Salem, in 1629, in *Bradford*, 266; the *Answer of the Boston Elders to Lechford*, p. 55, *post; Winthrop*, i. 31, 96, 114, 115, 116, 180. "Some difference there was" about the ordination of Rev. Thomas Carter, at Woburn, in 1642, — when "some advised, in regard they had no elder of their own, nor any members very fit to solemnize such an ordinance,

they would desire some of the elders of the other churches to have performed it; but others supposing it might be an occasion of introducing a dependency of churches, etc., and so a presbytery, would not allow it." (*Winthrop*, ii. 91.) A few years later, Hooker (*Survey*), ii. 59) maintained that, "though it be most comely that those of *the same* Congregation should exercise it [the act of ordination], yet the *Elders* also of *other* Congregations may be invited hereunto;" and the Cambridge Synod of 1649, " saw not why imposition of hands might not be performed by the elders of other churches." — *Platform*, c. ix. § 5. Comp. *Magnalia*, b. v. pt. 2.; *Hist. Remarks*, § 5. (fol. p. 42); and *Ratio Disciplinæ*, 14-42.

⁵ "Testifying their brotherly acceptance of him, . . . and doth exhort him in the Lord, to fulfil the ministry which he hath received of the Lord." — Cotton, *Way*, 42.

Newes from New-England.

Church, divers weeks before; and fo alfo of every ordination. And fome Minifters, or others, as Meffengers from other Churches, are ufually prefent at fuch gatherings of Churches, and ordinations: for fometimes, Magiftrates, Captains, Gentlemen, and other meaner Brethren, are made meffengers of Churches, for thofe and other purpofes, never having had impofition of hands: And at planting of a Church, or gather- | ing, as they tearme it, one of the Church meffengers of forraine Churches, examines and tries the men to be moulded into a Church, difcerns their faith and repentance, and their Covenant being before ready made, written, fubfcribed, and here read and acknowledged, hee decerns[6] and pronounceth them to be a true Church of Chrift, and gives them the right hand of fellowfhip, and all this in the name of Chrift, and of all the Church-meffengers prefent, and their Churches: fo did Mafter *Weld* at the founding of *Weymouth* Church, or to this effect.[7]

---

[6] Decides, determines, adjudges.— *Richardfon.* From the Latin, *decernere.*

[7] "The Churches indeed fend Meffengers (commonly their Elders) to lend them a word of counfell if they need, being more experienced in thofe ways then (commonly) new beginners are, to joyne their prayers with theirs, and to give them the right hand of fellowfhip. . . . The Meffengers [if unfatisfied with any of thofe that are about to enter into church fellowfhip,] never arrogated to themfelves fuch power, to this day, (nay, they profeffedly expreffe againft it, conftantly in fuch meetings) as *to forbid their entrance* into Church eftate. The moft they doe (at any time in this cafe) is, to defire leave to be faithfull in interpofing their counfell, and that only when they fee very great caufe: And

Plaine dealing,

And the generall Court will not allow of any Church otherwise gathered.[8]

Some differ. Some Ministers have there heretofore, as I have heard, disclaimed the power of their Ministery received in *England*, but others among them have not.[9] Generally, for

withall leave them to their Christian liberty." — Welde, *Reply to W. R.*, 34, 35. Comp. *Winthrop*, i. 183, 184. Concerning the "founding of Weymouth Church," which was regathered Jan. 30, 1639, "with approbation of the magistrates and elders," — see *Winthrop*, i. 287, 288; and p. 22, *post.*

[8] See order of March, 1636, *Mass. Records*, i. 168.

[9] There had been some differences of opinion among the ministers of Massachusetts as to the validity of episcopal ordination, — accordingly as they inclined more to non-conformity, or separate congregationalism; though all agreed in holding that "the essence of ministers' calling under the Gospel, is the Congregation's consent." The "Confession and Protestation of the Faith of Certain Christians in England," &c. (1616), attributed, though questionably, to Henry Jacob, was explicit on these points: "We cannot but believe it to be simply unlawful and sinful, to fetch, receive, yea, or *to use*, a ministry formerly received from the Prelates;" and "that a minister, so reputed, without any particular flock, is indeed no minister." (*Hanbury's Memorials*, i. 296.) So Robinson (*Justif. of Separation*, 334,) says, "The judgement ... of the most forward men in the Land, in this case, I may not omit; which is, that *they renounce, & disclaym their ordination by the Prelates*, and hold their Ministery by the peoples acceptation." But while Rev. George Phillips of Watertown (who came over with Winthrop,) had declared, that "if they will have him stand minister by that calling which he received from the prelates in England, he will leave them," — Mr. Wilson was constituted teacher of the church at Charlestown by imposition of hands, "with this protestation by all, that it was only as a sign of election and confirmation, not of any intent that Mr. Wilson should renounce his ministry he received in England." — Fuller's Letter to Gov. Bradford, 1 *Mass. Hist. Coll.*, iii. 74; *Winthrop*, i. 32.

In April, 1637, the ministers who met at Concord for the ordination of Mr. Bulkley and Mr. Jones, "resolved that such as had been ministers in England were lawful ministers by the call of the people, notwithstanding their acceptance of the call of the bishops, etc., (for which they humbled

the moft part, they hold the Paftors and Teachers offices to be diftinct; the Teacher to minifter a word of knowledg, the Paftor a word of wifdome, but fome hold them all one;[10] as in the Church of *Watertowne*, there are two

themfelves, acknowledging it their fin, etc.,) but *being come hither, they accounted themfelves no minifters*, until they were called to another church." *Winthrop*, 217, 218. Mr. Lenthall, who was called by the people of Weymouth, " ftanding upon his miniftery as of the Church of England, . . . was compelled to recant fome words," (as Lechford tells, p. 22); and at his examination by the Elders of the Bay, in January, 1639, Mr. Cotton faid to him that " his former ordination, not being given by them that had lawful power, and former election, will not ferve to make him a minifter here, except they [the people of Weymouth] were in a mutual covenant as a church before," &c. (*Mf. Notes by Robt. Keayne.*) But when Roger Williams cited the admiffion of Mr. Cotton "and others moft eminent in New England," that, "notwithftanding their former profeffion of miniftry in Old England, yea, in New England," " they were but private Chriftians, until they received a calling from a particular church," (*Mr. Cotton's Letter Examined*, 1644; *Bloudy Tenent*, ch. xxvii.), — Mr. Cotton replied, that, " being caft out by the ufurping power of the prelacy, and

difmiffed, though againft their wills," they did look upon themfelves "as *private members* and *not officers* to any church here," until called, &c.; but that any other fenfe given to their declaration was either a miftake or " a fraudulent expreffion" of their minds. — *Reply to Mr. Williams's Anfwer*, p. 131.

Hooker (*Survey*, ii. 50, 51,) declares that the doctrine of an " *indelebilis* character" impreffed by ordination, " comes out of the forge of Popery, and is fo befooted with the smoake of the bottomleffe pit, and carried along in the fogs of the myfteries of iniquity, that by a secret fleight it hath eaten infenfibly into the *orders of Chrift* before the world was aware."

[10] William Rathband, in his "Narration of the Opinions and Practifes of the Churches lately erected in New-England" (London, 1644.) afferted, that " whereas, in opinion and tenent they precifely diftinguifh between the paftor's and teacher's office, yet in practife they ufually confound them: both Paftour and Teacher equally teaching and equally applying both the Word and Seales, without any difference." (p. 42.) Thomas Welde, in " An Anfwer to W. R. his Narra-

Paſtors," neither will that Church ſend any meſſengers to any other Church-gathering or ordination.

How members are received or added to the Church there.

When a man or woman commeth to joyne unto the Church ſo gathered, he or ſhee commeth to the Elders in private, at one of their houſes, or ſome other place ap-

tion," &c. (printed the ſame year, at London,) declares this ſtatement untrue; "for it is both our profeſſed judgements and *conſtant practice*, that as the teacher is choſen, whoſe proper gift is aptneſſe to teach, ſo after hee is choſen, hee bends himſelfe that way, and waites upon teaching, ſo the Paſtor upon exhorting, as *Rom.* 12. 7, 8. Though in ſuch congregations where there is but one, hee labours to improve his talent both waies, for the preſent neceſſity, till that defect be ſupplyed:" and citing from Mr. Cotton's Catechiſm, p. 2, "The Paſtor's worke is to attend upon exhortation; The Teacher on Doctrine," — adds: "His owne, and others practiſes there run accordingly" (p. 57). — Comp. Hooker's *Survey*, ii. 19, 21; Savage, *Note on Winthrop*, i. 31; Dexter's *Congregationaliſm*, 125.

There was "a very ſharp debate anent the office of Doctors," (or Teachers,) in the Weſtminſter Aſſembly, in 1643. The Independents "were for the divine inſtitution of a Doctor [Teacher] in every congregation, as well as a Paſtor." The Presbyterians were "extremely oppoſite:" but a final agreement was had on certain propoſitions "wherein the abſolute neceſſity of a Doctor in every congregation, and his divine inſtitution, in formal terms, was eſchewed, yet where two miniſters can be had in one congregation, the one is allowed, according to his gift, to apply himſelf moſt to teaching, and the other to exhortation; according to the Scriptures." — *Baylie's Letters*, in *Hanbury*, ii. 217.

" George Phillips and John Knowles. Winthrop (ii. 18), when recording, under date of Dec. 9, 1640, the ordination of Mr. Knowles, "a godly man and a prime ſcholar," remarks: "And ſo they had now two paſtors and no teacher, differing from the practice of the other churches, as alſo they did in their privacy, not giving notice thereof to the neighboring churches, nor to the magiſtrates, as the common practice was."

A few weeks after Mr. Wilſon's return from England, in 1632, the Boſton Church ſought advice from the elders and brethren of Plymouth, Salem, etc., on the queſtion "Whether there might be divers paſtors in the ſame church?" to which the reſponſe was, "Doubtful." — *Winthrop*, i. 81.

pointed, upon the weeke dayes, and make knowne their
defire, to enter into Church-fellowfhip with that Church,
and then the ruling Elders, or one of them, require, | or
afke him or her, if he bee willing to make known unto
them the worke of grace upon their foules, or how God
hath beene dealing with them about their converfion:
which (at *Bofton*) the man declareth ufually ftanding, the
woman fitting. And if they fatisfie the Elders, and the
private affembly, (for divers of the Church, both men
and women, meet there ufually) that they are true beleevers, that they have beene wounded in their hearts for
their originall finne, and actuall tranfgreffions, and can
pitch upon fome promife of free grace in the Scripture,
for the ground of their faith, and that they finde their
hearts drawne to beleeve in Chrift Jefus, for their jufti-
fication and falvation, and thefe in the minifterie of the
Word, reading or conference: and that they know competently the fumme of Chriftian faith. And fometimes,
though they be not come to a full affurance of their good
eftate in Chrift. Then afterwards, in covenient time, in
the publique affembly of the Church, notice is given by
one of the ruling Elders, that fuch a man, or woman, by
name, defireth to enter into Church-fellowfhip with them,
and therefore if any know any thing, or matter of offence
againft them, for their unfitneffe to joyne with them, fuch
are required to bring notice thereof to the Elders; elfe,

5

The ufuall
termes where-
upon.

that any who know them, or can say any thing for their fitneffe, be ready to give teftimony thereof, when they fhall be called forth before the whole Church.

6

Matters of offence how heard in private.

If there be matter of offence, it is firft heard | before the Elders, and if the party fatisfie them, and the offended, in private, for private offences, and promife to fatisfie in publique, for publique offences; then, upon another day, one of the ruling Elders calleth forth the party, by name, in the publique affembly of the Church, and before ftrangers, and whomfoever prefent, moft commonly upon the Lords day, after evening exercifes, and fometimes upon a week day, when all the Church have notice to be prefent.

Dilatorie proceedings in admitting members.

The party appearing in the midft of the Affembly, or fome convenient place, the ruling Elder fpeaketh in this manner: Brethren of this congregation, this man, or woman *A. B.* hath beene heretofore propounded to you, defiring to enter into Church-fellowfhip with us, and we have not, fince that, heard any thing from any of you to the contrary, of the parties admittance, but that we may goe on to receive him: Therefore now, if any of you know any thing againft him, why he may not be admitted, you may yet fpeak. Then after fome filence he proceedeth, Seeing no man fpeaketh to the contrary of his admiffion, if any of you know any thing, to fpeak for his receiving, we defire you, give teftimony thereof to the

Church, as you were alſo formerly defired to be ready therewith, and expreſſe your ſelves as briefly as you may, and to as good hearing. Whereupon, ſometimes, men do ſpeak to the contrary, in caſe they have not heard of the propounding, and ſo ſtay the party for that time alſo, till this new offence be heard before the | Elders, ſo that ſometimes there is a ſpace of divers moneths between a parties firſt propounding and receiving; and ſome are ſo baſhfull, as that they chooſe rather to goe without the Communion, then undergoe such* publique confeſſions and tryals, but that is held their fault.[12]

7

* Whether Popiſh Auricular confeſſion, and theſe public confeſſions be not extremes, and whether ſome private Paſtorall or Presbyteriall collation, left at liberty, upon cauſe, and in caſe of trouble of conſcience, as in the Church of England is approved, be not better then thoſe extremes, I leave to the wife and learned to judge.

But when none ſpeaketh to the contrary, then ſome one, two, or three, or more of the Brethren ſpeak their

Teſtimonials and Recommendations.

---

[12] Comp. Cotton's *Way*, pp. 53-55. "In this trial," he ſays, "we do not exact eminent meaſure, either of knowledge, or holineſs, ... for we had rather ninety-nine hypocrites ſhould periſh through preſumption, than one humble ſoul belonging to Chriſt ſhould ſink under diſcouragement or diſpair." (p. 58.) Yet Mather (*Magnalia*, b. v. pt. ii. 43, 44,) commenting upon certain "difficulties" in the platform of diſcipline, corroborates Lechford's ſtatement. "The Jews tell us of כלבא. or a *Scare-Crow* upon the top of the *Temple*, which kept off the *fowls* from defiling of it; and it hath been the Opinion of many that this *Cuſtom* of *Relations*, to be made by Candidates for *Admiſſion* to the Church . . . is as a Scare-Crow to keep Men out of the *Temple;* but, it may be, it has been the Opinion of as many, that none but the *Defilers* of the *Temple* would be kept out by ſuch a *Scare-Crow*. . . . Well, the reſult of theſe various Apprehenſions has been this: That ſome *unſcriptural Severities*, urged in this matter by ſeveral of our Churches, in the beginning of the Plantation, are now generally laid aſide," etc. So, Increaſe Mather (in the Epiſtle prefixed

opinions of the party, giving inſtances in ſome godlineſſe and good converſation of his, or ſome other recommendation is made, and that they are willing (if the Church thereto conſent) for their part, to give him the right hand of fellowſhip.

Which done, the Elder turneth his ſpeech to the party to be admitted, and requireth him, or ſometimes aſketh him, if he be willing to make knowne to the congregation the work of grace upon his ſoule; and biddeth him, as briefly, and audibly, to as good hearing as he can, to doe the ſame.

Publique confeſſions of parties to be received.

Thereupon the party, if it be a man, ſpeaketh himſelfe; but if it be a woman, her confeſſion made before the Elders, in private, is moſt uſually (in *Boſton* church) to the Life of Mitchell,) ſays, "It cannot be denied ... that there has been an unjuſtifiable *Severity* in impoſing *Circumſtantials* not inſtituted, whereby ſome truly gracious Souls have been diſcouraged from Offering themſelves to joyn in Fellowſhip with ſuch Churches. Thus it hath been, when an Oral Declaration of *Faith* and *Repentance* has been enjoyned on all Communicants, and that before the *whole Congregation;* when as many an Humble Pious Soul has not been gifted with ſuch *Confidence*." — (*Magnalia*, b. iv. c. 4. p. 159.)

Mr. Hooker ſpoke more pointedly. After laying down the rule, that "if a perſon live not in the commiſſion of any known ſin, nor in the neglect of any known duty, and can give a reaſon of his hope towards God," he is to be judged fit for church-ſociety, — he remarked (*Survey*, iii. 6), "This rule being received and agreed upon, it would mervailouſly facilitate the work of *Admiſſion*, without any trouble, and prevent ſuch *curious inquiſitions and niceties, which the pride and wantonneſſe of mens ſpirits hath brought into the Church*, to diſturb the peace thereof, and to prejudice the progreſſe of God's Ordinances." [Comp. *Cambridge Platform*, c. xii. § 3.]

read by the Paſtor, who regiſtred the fame.[13] At *Salem* the women ſpeake themſelves, for the moſt part, in the Church; but of late it is ſaid, they doe this upon the week dayes there, and nothing is done on Sunday, but their entrance into Covenant. The man in a ſolemne ſpeech, ſometimes a quarter of an houre long, ſhorter or longer, declareth the work of grace in his ſoule, to the fame purpoſe, as that before the Elders formerly mentioned. 8

Then the Elder requireth the party to make profeſſion of his faith; which alſo is done either by queſtions and anſwers, if the party be weake, or elſe in a ſolemne ſpeech according to the ſumme and tenour of the Chriſtian faith laid downe in the Scriptures, defining faith, and ſhewing how it is wrought by the Word, and Spirit of God, defining a Church to be a company of beleevers gathered out of the world, by the Word preached, and holy Spirit, and knit together by an holy Covenant, that there are in the Church remaining ſuch and ſuch officers,

Their profeſſion of faith.

---

[13] "In the churches where we have lived many years, we have ſeene ſuch a tender reſpect had to the weaker ſex (who are uſually more fearefull and baſhfull) that we commit their triall to the Elders and ſome few others in private, who upon their teſtimony are admitted into the Church without any more adoe."—T. Welde, *Anſwer to W. R.*, 19. "Some, being more weake and fearefull," ſays the fame writer (p. 48,) "we rather tender (as Jacob would not overdrive the feabler ſort of ewes and lambes) leſt they ſhould miſcarry."—Comp. Hooker's *Survey*, iii. 6; *Cambr. Platform*, ch. xii. § 4.

Officers in the Church.

and members, as aforesaid: That is to say, Pastors and Teachers, ruling Elders, Deacons and Deaconesses or Widowes;[14] and such and such are their offices and

Their duties or offices.

duties in particular, *viz.* the Pastor to exhort, and besides to rule; the Teacher to instruct in knowledge, and likewise to rule; the ruling Elder[15] to assist Pastor and

---

[14] See, after, p. 15,—"No Church there hath a Deaconnesse, as far as I know." Robert Browne (*The Points and Parts of all Divinity.* Middleburgh, 1582,) names the Widow, as "a person having office of God to pray for the Church, and to visit and minister to those which are afflicted and distressed in the Church." *Defin.* 54. (*Hanbury's Memorials*, i. 21.) "Their Relievers, or Widows, must be women of sixty years of age at the least, for avoiding of inconveniences," &c.— *True Descrip. of the Visible Church*, 1589. (*Ibid.* 30.) Comp. J. Canne's *Necessitie of Separation*, 6. Gov. Bradford mentions "one ancient widow . . . a Deaconnefs," in his time, in the church at Amsterdam.— *Dialogue between some Young Men*, &c., Young's *Chron. of Plymouth*, 455.

Mr. Cotton regarded Widows as "fit assistants to the Deacons, in ministering to the sick," etc., . . . "onely we find it somewhat rare to find a woman of so great an age . . . fit to undertake such a service."— *Way of the Congreg. Churches*, p. 39. Comp. *Cambr. Platform*, c. vii. § 7.—"The Lord hath appointed ancient widows, I *Tim.* v. 9, 10, (where they may be had,) to minister" etc. Mr. Davenport (*Catechism;* repr. N. Haven, 53,) names four officers of "the second sort of ministry : . . . 4. The Deacon, . . . under whom is included the widow or Deaconefs, who is to attend the sick and impotent," &c.

[15] Savage, on Winthrop, i. 31, note 3. For ample citations of early authorities, and a history of this office from its origin to its decline, and, finally, its entire disuse, in Congregational churches, see Rev. Dr. Dexter's *Congregationalism*, pp. 110–132. "The latest record on the books of the First Church in Boston, of the election of a Ruling Elder is believed to be of date, August 3, 1701."— *Ibid.* 131. A few years earlier, Joshua Scottow lamented that, while "some of the Old Planters children" remembered "that there were such men, when they were young, that were called Ruling Elders, . . . what men they were, or what was their work, they professed they could not tell."— *Narrat. of the Planting*, &c. (1694), in 4 *Mass. Hist. Coll.*, iv. 329.

Teacher in ruling, as the Levites were given to the
Priefts for helps, and to fee to whomfoever comming into, or to goe forth of the Church, by admonition,[16] or
excommunication; the Deacon to receive the contributions of the Church, and faithfully to difpofe the fame;
the Deaconeffes to fhew mercie with cheerfulneffe, and
to minifter to the fick and poore brethren; the members <small>Members duties.</small>
all, to | watch over and fupport one an other in broth- 9
erly love.

Notwithftanding, there was a Sermon lately made by <small>A Sermon of</small>
Mafter *Cotton* in *October, Anno* 1640. upon 1 *Cor.* 11. 19.
touching herefies, which was fince commonly there called
the Sermon of the twelve Articles, wherein was declared,
that there are twelve Articles of Religion, which maintained by any, the Church may receive them, and keepe
fellowfhip with them; but the ignorant[17] of them after
inftruction and fcandalous fins unrepented, exclude from
the fellowfhip of the Church. The faid Articles were
to this effect: Firft, that there are three Perfons in one <small>twelve Articles of Religion.</small>
God, the Father, the Sonne, and the holy Spirit. Sec-
ondly, that this God made, and governs all the World,
and that he is a rewarder of the good, and punifher of
the evill. Thirdly, that this God alone is to be wor-
fhiped. Fourthly, this worfhip of God is inftituted in

<small>16 The Mafs. Hift. Society's Ms. has "admiffion."   17 The fame Ms. has "ignorance," for "ignorant."</small>

his written Word, not the precepts of men. Fiftly, that from the fall of *Adam*, we have not so worshiped God, but have all sinned, and deprived our selves of the reward promised, and therefore are under the curse by nature. Sixthly, that we are by nature utterly unable to rescue our selves from this curse. Seventhly, that Jesus Christ the eternall Sonne of God, in fulnesse of time took upon him our nature, and was made flesh for us, and by his death and sufferings, redeemed his elect from sin, and death. Eighthly, that Christ Jesus, and salvation by him, is offered, and given in the | Gospell, unto every one that beleeveth in his name, and onely by such received. Ninthly, that no man can come unto Christ, nor beleeve on him, except the Father draw him by his Word and Spirit. Tenthly, whom the Lord draws to him by his Word and Spirit, them he justifies freely by his grace and according to his truth, not by works. Eleventhly, where the soule is justified, it is also regenerate and sanctified. Twelfthly, this regeneration and sanctification is still imperfect in this life. And unto all is added this generall Article, That such as walke after this rule, shall arise to everlasting life; and those that walk otherwise, shall arise to everlasting condemnation, in the day of Judgement: That the knowledge and beliefe of these are of the *foundation of Religion:* But things touching the *foundation of Churches*, as Baptisme, Imposition of hands;

Newes from New-England. 27

ignorance in thefe may hinder the meafure of our reward in heaven, not communion with the Church on earth.[18] Exceptions againft the Apoftles Creed were thefe: That it is not of neceffity to beleeve Chrifts defcent into hell in any fenfe;[19] That it is not in that Creed contained,

[18] "Now, in points of doctrine fome are fundamental, without right belief whereof a man cannot be faved; others are circumftantial or lefs principal, wherein men may differ in judgment without prejudice of falvation on either part." Cotton's *Anfwer to Arguments againft Perfecution*, etc. To this diftinction, Roger Williams objected, believing that "God's people may err from the very fundamentals of vifible worfhip," and yet be faved. *Bloudy Tenent*, ch. iv. In his Reply (*Bl. Tenent Wafhed*, etc., p. 5) Mr. Cotton explains, that "fundamental doctrines are of two forts; fome hold forth the foundation of Chriftian religion — others concern the foundation of the Church:" and that he had fpoken, as above, "of the former fort of thefe only — the other fort I look at as lefs than principal, in comparifon with thefe." — *Hans. Knolly's Soc. ed.*, pp. 19, 39.

[19] The controversy on this article of belief was "plied hotly in both the univerfities, in 1604, and after," when Mr. Cotton was at Cambridge. — Wood's *Athenæ. Oxon.* (ed. Blifs), ii. 308. Certain fermons preached at St. Paul's Crofs, London, in 1597, by Bilfon, Bifhop of Winchefter, in which the doctrine of Chrift's defcent to hell was maintained, had given much offence to the Puritans; and the next year Henry Jacob publifhed "A Treatife of the Sufferings and Victory of Chrift, . . . declaring by the Scriptures . . . That Chrift after his death on the Crofs, went not into Hell in his Soule; contrarie to certaine Errours in thefe points publickly preached in London." (1598, 8vo. pp. 174.) "The Effect of certain Sermons, touching the full Redemption of Mankind by the Death and Bloud of Chrift Jefus," etc., was printed by Bifhop Bilfon, in 1599 (Lond. 4to.), and anfwered by Jacob, in "A Defence of a Treatife," etc. (1600, 4to. pp. 211.) At the fuggeftion of Queen Elizabeth, as is ftated, the bifhop prepared a more full and elaborate defence of his fermons, and of the doctrine in controverfy, in "The Survey of Chrift's Sufferings for Man's Redemption and of his Defcent to Hades or Hell," etc. (Lond. 1604, fol.) — Wood's *Athen. Oxon.*, ut fupra, and ii. 170, 171, 309; Hanbury's *Memorials*, i. 221. Robert Parker publifhed, in refutation of Bilfon, and other affert-

that the Scripture is the onely rule of Gods worſhip; nor doth it ſo directly ſet forth the point of Juſtification.

<small>Maſter *Knolls* how admitted.</small>

And alſo I remember Maſter *Knolles*,[20] now one of the Paſtors at *Watertowne*, when he firſt came to be admitted at *Boſton*, never made any mention in his profeſſion of faith, of any Officers of the Church in particular, or their duties, and yet was received.

<small>11
Right hand of fellowſhip given to brethren.</small>

The party having finiſhed his Diſcourſes of his confeſſion, and profeſſion of his faith, the Elder againe ſpeaketh to the congregation: Brethren of the congregation, if what you have heard of, [and][21] from this party, doe not ſatisfie you, as to move you to give him the *right hand of fellowſhip*, uſe your liberty, and declare your mindes therein: And then, after ſome ſilence, if none

<small>The whole Church ruleth.</small>

except againſt the parties expreſſions, (as often ſome members doe) then the Elder proceedeth, ſaying, But if you are ſatisfied with that you have heard of, and from him, expreſſe your willingneſſe, and conſent to receive him, by your *uſuall ſigne*, which is *erection and extention of the right hand*.[22]

<small>ers of this doctrine, "De deſcenſus Domini noſtri, Jeſu Chriſti ad Inferos, libri quatuor, ab Hugoni Sanfordo inchoati, opera R. P. ad umbilicum perducti." (Amſt. 1611, 4to.)

[20] Rev. John Knowles had been a fellow of Catharine Hall, Cambridge. He was admitted to the Boſton Church, Aug. 15, 1639, and was ordained at Watertown, Dec. 9, 1640. — Savage, *Geneal. Dict.; Winthrop*, ii. 18.

[21] The conjunction is inſerted on the authority of the M. H. S. Ms.

[22] See after, p. 12, and note 25; p. 14, note 37.</small>

This done, sometimes they proceede to admit more members, all after the same manner, for the most part, two, three, foure, or five, or more together, as they have time, spending sometimes almost a whole afternoone therein. And then the Elder calleth all them, that are to be admitted, by name, and rehearseth the covenant, on their parts, to them, which they publiquely say,[23] they doe promise, by the helpe of God, to performe: And then the Elder, in the name of the Church, promiseth the Churches part of the covenant, to the new admitted members. So they are received, or admitted. *Their enterance into Covenant.*

Then they may receive the Sacrament of the Lords supper with them, and their children bee baptized, but not before: also till then they may not be free men of the Common-wealth, but being received in the Church they may.

Sometimes the Master is admitted, and not the servant, & e contra: the husband is received, and not the wife; and on the contrary, the child, and not the parent. *12 Severing in the family.*

Also all matters of publique offence are heard & determined in publique, before all the Church, (and strangers *Offences, how heard in publique.*

---

[23] Mr. Welde (Answer to W. R., 24) writes: "He [Rathband] tells us, *We hold our Church Covenant must be vocall.* . . . It's contrary (wee are sure) to our constant practise, that admits members into the Church by a Covenant agreed to by *their silence only:* and as it is contrary to our practise, so to our writing, in the *Discourse of the Covenant,* which expressly saith, *that silent consent is sufficient.*"

Plaine dealing,

<small>The whole Church ruling and ufurping the keyes.

* Whether a grave and judicious confiftorie of the Bifhop well affifted be not a great deale better, I leave to our fuperiours to determine.</small>

too in *Bofton*,²⁴ not fo in other places.) The party is called forth, and the matter declared and teftified by two witneffes; then he is put to anfwer: Which finifhed, one of the ruling Elders afketh the *congregation if they are fatisfied with the parties expreffions? If they are, he requireth them to ufe their *liberty*, and declare their fatiffiedneffe; If not, and that they hold the party worthy of admonition or excommunication, that they witneffe their affent thereto by their filence.²⁵ If they be filent, the fentence is denounced. If it be for defaults in erroneous opinions onely, the Teacher, they fay, is to denounce

<small>²⁴ "Some of our moft populous Churches do no Church Act, no not of difcipline, but in the prefence of the whole Towne, (non-members, as well as members) fo many of them as are pleafed to be prefent. Wayes of truth feeke no corners; if any Church admonifh a brother privately, it is becaufe his offence is not known to non-members." — Cotton, *Way cleared*, pt. i. p. 68.

²⁵ "The whole Church may be faid to bind and loofe, in that the Brethren confent and concurre with the Elders, both before the Cenfure, in difcerning it to be juft and equall, and in declaring their difcernment, by lifting up of their hands, or by filence," etc. — Cotton, *Keyes*, 14. "The confent of the people gives a caufall vertue to the compleating of the fentence of excommunication." — Hooker, *Pref. to Survey*. "Its granted by Divines, there can be no proceeding to *excommunication*, but with the *tacite confent of the people*." — *Survey*, pt. i. p. 135. Comp. Cotton, *Way*, 92; *Cambr. Platform*, c. x. §§ 5, 9, 10.

A memorial prefented to the court at Ipfwich, by certain members of the Newbury Church, in 1669, fays: "Near thirty years fince, at a fynod at Cambridge, it was propofed, and it was confented unto by them, that if the minifters thought it moft convenient to vote by fpeech and filence, rather than by lifting up the hand, they had nothing againft it, feeing the one was a teftimony of confent as well as the other, fo this kind of voting began and continued in practice without difference or interruption for a good feafon." — Coffin's *Newbury*, 78.</small>

Newes from New-England.

the fentence; If for matter of ill manners, the Paftor de- _{Who denounce Church cenfures.}
nounceth it; the ruling Elders doe not ufually denounce
any fentence:[26] But I have heard, a Captaine[27] delivered
one to Satan, in the Church at *Dorchefter*, in the abfence
of their Minifter.

Ordinarily, matter of offence is to be brought to the _{*Dic Ecclefiæ.*}
Elders in private, they may not otherwife *tell the* *Church* _{* This agreeth with the rule in England.}
in ordinary matters, and fo it hath been declared in pub-
lique, by the Paftors[28] of *Bofton*.[29]

The admonifhed muft, in good manners, abftain from _{Admonition.}
the Communion, and muft goe on to fatisfie the Church,
elfe Excommunication follows.

[26] To the contrary, Hooker (*Survey*, iii. 38,) lays down the rule, that, after the affent of the Church has been given, "the fentence, thus compleatly iffued, is to be folemnly paffed and pronounced upon the Delinquent, *by the ruling Elder*, whether it be the cenfure of *admonition* or *excommunication*." Cotton (*Keyes*, 22) does not difcriminate, but gives to "the Elders" authority "both *jus dicere*, and *fententiam ferre*." So, the Cambridge Platform, c. vii. § 2, included with thofe acts of fpiritual rule in which the Ruling Elders are to join with the Paftor and Teacher, that of pronouncing fentence. Comp. Cotton's *Way*, 91, 92. Winthrop, in his mention of Mrs. Hutchinfon's excommunication, fays that "it being for manifeft evil in matters of converfation ... the fentence was denounced by the paftor [Mr. Wilfon], *matter of manners belonging properly to his place*." (i. 258.)

[27] Whofe name, "Ifrael Stoughton," is given in the Maffachufetts Hiftorical Society's Ms.

[28] "Paftor." — *Mafs. Hift. Soc. Ms.*

[29] ... "The brother firft offended telleth the church of it, to wit, in God's way: he telleth the elders, who are the mouth of the church," etc. — Cotton, *Way*, 90. "When there be Elders in a Church, all the complaints *muft* be made to them, and the caufes prepared and cleared, and then by their means they muft be complained of to the Church." — Hooker, *Survey*, i. 134, 135; fo iii. 36.

13
Excommunication.

The excommunicate is held *as an Heathen and Publican:* Yet it hath been declared at *Boston* in divers cases, that children may eate with their parents excommunicate;[30] that an elected Magistrate excommunicate may hold his place, but better another were chosen;[31] that an

---

[30] Such a declaration had been made by Mr. Wilson, after the excommunication of Mrs. Hutchinson: "In the general, he said indeed, that with excommunicate persons no religious communion is to be held, nor any civil familiar connexion, as fitting at table. But ... such as were joined in natural or civil near relations, as parents and children, husband and wife, &c., God did allow them that liberty, which he denies to others." — Cotton's letter to Fras. Hutchinson, in 2 *Mass. Hist. Coll.* x. 186. Comp. Cotton, *Way,* 93, 94; Hooker, *Survey,* iii. 39; *Cambr. Platform,* c. xiv. § 5; S. Mather, *Apology,* 108.

[31] "Excommunication ... toucheth not princes or magistrates in respect of their civil dignity or authority." — *Cambr. Platform,* c. xiv. § 6.

No civil disabilities followed excommunication except disqualification for admission as a freeman. In England, even so late as the 53d of George III. (1813) the excommunicate was debarred from serving as a juryman, from bringing or maintaining actions, from appearing as a witness in any cause, from practising as an attorney in any court; and from doing any act "that is required to be done by one that is *probus et legalis homo.*" The excommunicate was moreover liable, after forty days, to be taken on writ *de excommunicato capiendo* (issued on the bishop's certificate), and to be imprisoned in the county jail, till he should be reconciled to the church. — Blackstone, *Comment.* iii. 102.

For a single year Massachusetts had a law that any person who should "stand excommunicate for the space of six months, without labouring what in him or her lyeth to bee restored," should be presented to the Court of Assistants, and proceeded with "by fine, imprisonment, banishment, or *further,* for the good behaviour, as their contempt and obstinacy, upon full hearing, shall deserve." — *Mass. Rec.,* i. 242. This law was enacted in September, 1638, and repealed September, 1639. — *Ibid.* 271.

Roger Williams (in *The Bloudy Tenent,* c. cxxviii.) mentions this "strange law in New England formerly," by way of explaining a supposed reference to it in "A Model of Church and Civil Power," &c., the authorship of which he mistakenly as-

hereditary Magiftrate, though excommunicate, is to be obeyed ftill in civill things; that the excommunicate perfon may come and heare the Word, and be prefent at Prayer, fo that he give not publique offence, by taking up an eminent place in the Affembly: But at *New-haven, alias Quinapeag,* where Mafter *Davenport* is Paftor, the excommunicate is held out of the meeting, at the doore, if he will heare, in froft, fnow, and raine.[32]

cribed to Mr. Cotton (fee *Bloody Tenent wafhed,* etc., pp. 150, 192): "To give liberty to Magiftrates, without exception, to punifh all excommunicate perfons within fo many months, may" (fay the writers of the Model) "prove injurious to the perfon who needs, to the church who may defire, and to God who calls for longer indulgence from them." Mr. Cotton's opinions on this fubject may be found in his *Expofition upon Revelation,* c. xiii. (delivered, January–April, 1640): "It was a matter in queftion here not long agoe, whether the Court fhould not take a courfe to punifh fuch perfons as ftood excommunicate out of the Church, if they fhould ftand long excommunicate, but it was a good providence of God that fuch a thing was prevented: Let not any Court, *ipfo facto,* take things from the Church." (p. 19.) Again, "It is dangerous to bring in civill Authority immediately upon Church-cenfure: A warning to us

here, that if men be excommunicated, not to deny them civill Commerce, or to fay fuch as ftand out excommunicated fo long, fhall no longer enjoy the priviledges of the State." (*Ibid.,* p. 238.)

[32] On this, Dr. Bacon (in *Hiftorical Difcourfes,* 48) remarks: "Lechford was probably lawyer enough to know that the fame rule obtained in the Church of England, and that the excommunicate, befides being excluded from the place of worfhip, was liable to a penalty every Sunday for his confrained abfence. Good old Oliver Heywood found that this was no dead letter. *Heywood's Works,* i. 100." — See the Acts of 1 Eliz., c. 2; 23 Eliz., c. 1 (impofing fines on every abfentee from the parifh church); and 7 Jac. 1. c. 6; *Blackftone's Comment.,* iv. 52. One of the fchifmatical tenets for maintaining which feveral non-conformifts of Northamptonfhire were called to anfwer Laud's Ecclefiaftical Commiffiioners, in 1634, was, "that

Moft an-end, in the *Bay*, they ufe good moderation, and forbearance in their cenfures: Yet I have known a Gentlewoman excommunicate, for fome indifcreet words, with fome ftifneffe maintained, faying, A brother, and others, fhe feared, did confpire to arbitrate the price of Joyners worke of a chamber too high, and endeavouring to bring the fame into civill cognizance, not proceeding to take two or three to convince the party, and fo to tell the Church, (though fhee firft told the party of it) and this without her husband. I feare fhe is not yet abfolved; I am fure fhe was not upon the third of Auguft laft, when we loofed from *Bofton*.

Cognizance of caufes.

There hath been fome difference about jurifdictions, or cognizance of caufes: Some have held, that in caufes betweene brethren of the Church, the matter fhould be

perfons excommunicated by the ordinary, might come to church."— *Calendar of Brit. State Papers*, 1634-35, p. 411.

In 1644, Henry Glover, who had been excommunicated by the Church of New Haven, expreffed a defire to be reftored. "The brethren agreed that he fhould have liberty to fpeak in the afternoon," when, after the contribution was ended, "the ruling elder defired *fome that ftood near the door, to call in* Henry Glover." Mr. Davenport then addreffed him, telling him of the law in Leviticus xiii. and xiv., concerning the cleanfing of the leper, and explained how "the leper under the law anfwered the ftate of an excommunicated perfon now."— *N. H. Church Rec.*, in Bacon's *Hift. Difcourfes*, 307-309. See, too, in reference to Mrs. Eaton's cafe, in Trial of Ezek. Cheever, *Coll. Conn. Hiftor. Society*, i. 29, 44.

The church at Bofton did not debar the excommunicate from entrance into the affembly, "in time of preaching the Word, or Prayer, or fuch other worfhip of God as is not peculiar to the church; for this liberty we do not forbid to Heathens and Indians."— Cotton's *Way*, 93.

firſt told the | Church, before they goe to the civill Magiſtrate, becauſe all cauſes in difference doe amount, one way or other, to a matter of offence; and that all criminall matters concerning Church members, ſhould be firſt heard by the Church. But theſe opinioniſts are held, by the wiſer ſort, not to know the dangerous iſſues and conſequences of ſuch tenets.[33] The Magiſtrates, and Church-leaders, labour for a juſt and equall correſpondence in juriſdictions, not to intrench one on the other, neither the civill Magiſtrates to be exempt from Eccleſiaſticall cenſure, nor the Miniſters from Civill;[34] & whether Ec-

[33] Anthony Stoddard, one of the conſtables of Boſton in 1641, was one of theſe "opinioniſts," as appears from *Winthrop*, ii. 39, 40. When required by Gov. Bellingham to take in cuſtody Francis Hutchinſon, he "ſaid withal to the governour, Sir, I came to obſerve what you did, that if you ſhould proceed with a brother other-wiſe than you ought, *I might deal with you in a church way;*" and having been committed, for this "inſolent behavior," he admitted his error, "which was that he did conceive that the magiſtrate ought not to deal with a member of the church before the church had proceeded with him."

[34] The General Court, Sept. 1639, propoſing to take meaſures for the "preſent reformation of immoderate great ſleeves, and ſome other ſuperfluities" of apparel, found "that ſome [had] been grieved that ſuch exceſſes were preſented to the Court, which concerned the members of churches, before the parties had been dealt with at home," etc.; and thereupon, all proceedings upon ſuch preſentments were ſtayed, "in expectation that the officers and members of all the churches, having now clear knowledge . . . will ſpeedily and effectually proceed againſt all offenders in this kind, and . . . keep the more ſtrict watch . . . for time to come."—*Maſs. Rec.*, i. 274.

In October, 1640, the elders renewed a motion which had been made at a previous Court, "that the churches might know their power and the civil magiſtrate his. The ſame had been moved by the magiſtrates formerly, and now at this Court they preſented a writing to that effect,

clesiasticall, or Civill power first begin to lay hold of a man, the same to proceed, not barring the other to intermeddle.

Churches independent.
Every Church hath power of government in, and by it selfe; and no Church, or Officers, have power over one another but by way of advice or counsaile, voluntarily given or besought,[35] saving that the generall Court, now

to be considered by the Court, wherein they declared that the civil magistrate should not proceed against a church member before the church had dealt with him, with some other restraints which the Court did not allow of. So the matter was referred to further consideration, and it appeared indeed that divers of the elders did not agree in those points." — *Winthrop*, ii. 16, 17.

The history of this movement, and its influence in shaping and in securing the adoption of the "Body of Liberties," of 1641, deserve more thorough examination than they appear hitherto to have received from historians. No more difficult problem was presented to the founders of Massachusetts, than that of defining the limits of jurisdiction between the civil magistracy and the churches. "It is necessary," taught Mr. Cotton, at one of his weekly lectures, early in 1640, (after the body of laws, drawn up by a committee of the General Court had been sent to the elders and freemen of the several towns, for their consid-

eration,) "It is necessary ... that all power that is on earth be limited, church-power or other. ... It is counted a matter of danger to the State to limit Prerogatives; but it is a further danger, not to have them limited: They will be like a Tempest, if they be not limited. ... It is therefore fit for every man to be studious of the bounds which the Lord hath set: and for the People, in whom fundamentally all power lyes, to give as much power as God in his word gives to men: And it is meet that Magistrates in the Common-wealth, and so Officers in Churches should desire to know the utmost bounds of their own power, and it is safe for both;" etc.— *Expos. of 13th Chap. of Revelation*, 72.

"A Declaration of the Liberties the Lord Jesus hath given to the Churches," (comprising eleven articles,) was incorporated in the Body of Liberties established in 1641, — for which see 3 *Mass. Hist. Coll.*, viii. 234.

[35] "All particular Churches and all the Elders of them are of equal

Newes from New-England.

and then, over-rule fome Church matters: and of late, divers of the Minifterie have had fet meetings to order Church matters; whereby it is conceived they bend towards Presbyterian rule.[36]

power, each of them refpectively in their own Congregation. None of them call others Rabbies, or Mafters, or Fathers (in refpect of any authoritie over them) but all of them own and acknowledge one another as fellow brethren, Matth. 23. 8, 9, 10." — Cotton, *Keyes*, p. 37. Comp. *Way cleared*, ii. 20, 21; Hooker, *Survey*, i. 219, 220; *Cambr. Platform*, c. xv. § 1.

"Beware of all fecular power, and Lordly power; of fuch vaft infpection of one church over another: ... Leave every church Independent; not Independent from *brotherly counfell;* God forbid that we fhould refufe *that;* but when it comes to *power*, that one Church fhall have power over the reft, then look for a Beaft [Revel. xiii. 2], which the Lord would have all his people to abhor." — Cotton, *Expos. of Revel.* xiii. 30, 31.

"At all times, when a particular church fhall wander out of the way, (whether out of the way of truth, or of peace) the community of churches may by no means be excufed from reforming them again into their right way." — Cotton, *Keyes*, 59.

[36] The laft three lines of this paragraph, beginning "and of late," etc., are not in the M. H. S. Ms. — The "fet meetings" of the minifters had, from the firft, given offence to fome who held to the abfolute independence of the churches. In 1633, when "the minifters in the Bay and Sagus [Lynn] did meet, once a fortnight, at one of their houfes by courfe, when fome queftion of moment was debated," — the Salem paftor, Mr. Skelton, a rigid feparatift, and Roger Williams (then lately returned from Plymouth, and "exercifing by way of prophecy" at Salem, though not in church-office), "took fome exception" to thefe meetings, "as fearing it might grow in time to a presbytery or fuperintendency to the prejudice of the Churches' liberties." — *Winthrop*, i. 116, 117. "Mr. Williams [before his banifhment] had fome fellowfhip with us," faid Mr. Cotton (*Way cleared*, i. 55), "and might have had more, but that hee fufpected all the *Statos conventus* of the Elders to bee unwarrantable, and fuch as might in time make way to a Presbyteriall government."

The "Model of Church and Civil Power," drawn up about 1635, and which appears to have had the approval of Mr. Cotton (fee before, p. 13, note 31), propofes, "as the means appointed by God whereby he may me-

Difference of rule in Churches.

In *Boston*, they rule, moſt an-end, by unanimous conſent, if they can, both in admiſſions, and cenſures, and other things. In *Salem*, they rule by the major part of the Church: You that are ſo minded hold up your hands; you that are otherwiſe minded, hold up yours.³⁷

diately reform matters amiſs in our churches," meetings, "1. Monthly of ſome of the *elders* and *meſſengers* of the churches ... which are neareſt together, and ſo may moſt conveniently aſſemble together; ... [who may] conſult of ſuch things as make for the good of the churches.... 2. Annual, of all the *meſſengers* and *elders* of the churches ... ſometimes at one church, ſometimes at another, ... [to which] let all the churches ſend their weighty queſtions and caſes, ſix weeks or a month before the ſet time." Theſe aſſemblies, monthly and annual, were to "*do nothing by authority*, but only by counſel." — *Bloudy Tenent*, ch. cxxix. [*Hans. Knollys Soc.*, 1848, pp. 334–6]. In this plan, Roger Williams found "a moſt four and uncomely deformed look of a mere human invention," and denies that "general arguments from the plauſible pretence of Chriſtian fellowſhip, God's glory, &c., prove ſuch particular ways of glorifying God, without ſome precept or precedent of ſuch a kind." — *Ibid.* c. cxxx–cxxxiv.

The 7th Article of the Declaration of Liberties of the Churches, adopted with the Body of Liberties in December, 1641, as the fundamental law of the colony, ſecures to the Elders "free libertie to meete monthly, quarterly, or otherwiſe, in convenient numbers and places, for conferences and conſultations about Chriſtian and Church queſtions and occaſions." And the 11th Article allows and ratifies "as a lawfull libertie of the Churches," monthly meetings of the elders and any other of the brethren, of neighbouring churches, for "publique Chriſtian Conference about the diſcuſſing and reſolveing of ... doubts and caſes of conſcience concerning matter of doctrine or worſhip," ... but "onely by way of brotherly conference and conſultations." (*Body of Liberties*, 95 (7, 11); 3 *Maſs. Hiſt. Coll.*, viii. 234, 235.) The Synod at Cambridge, in June, 1643, agreed, "That Conſociation of churches, in way of more general meetings, yearly; and more privately, monthly, or quarterly; *as conſultative Synods;* are very comfortable, and neceſſary for the peace and good of the churches." — *Letter from N. E.*, quoted in *Reply of Two Brethren to A. S.* (Lond. 1644), p. 7. See *Hanbury*, ii. 343.

³⁷ See before, p. 11, and p. 12, note

In *Boston*, when they cannot agree in a matter, they will sometimes referre it to some select brethren | to   15

Confiftory. A better Confift-

25. "Whether matters be carried amongſt them by moſt voices or no, is not ſo generally agreed upon. Some affirme that the major part carries it againſt the leſſer part, yea, though the officers be in this leſſer part, and to ſhew ſtrong reaſon to the contrary... Others, that the whole body muſt agree, elſe nothing proceeds... Some, that things are not carried by voyces at all, but by truth and right, and according to God.... Sometimes they grant indeed all things are carried by conſent of all; but they explain it thus, *viz*.... If the leſſer party diſſenting neither can give ſatisfaction to the greater, nor will receive ſatiſfaction from them, but ſtill perſiſt in diſſenting, then doe the major part (after due forbearance, and calling in the counſel of ſome neighbouring churches) judicially admoniſh them; who being thus *under cenſure*, their voyce is now extinct, and made voide. And ſo the reſt proceed to vote," etc. —[W. Rathband's] *Brief Narration of ſome Church Courſes in N. E.*, 27, 28. Comp. *Anſw. to the* 32 *Queſtions* [by Richard Mather], 58, 61. "When we ſay we do this or that with *common conſent*, our meaning is, wee do not carry on matters either by the *over-ruling power of the Preſbytery*, or by the conſent of the *major part* of the church; but by the generall and joynt conſent of all the members ... ὁμοθυμαδὸν, that is, *with one accord*, Acts 2. 46, as becometh the church of God." — Cotton, *Way*, 94. [The expedient of putting a diſſenting minority under cenſure, by admonition, and thereby nullifying their vote, was reſorted to in the Boſton church, in the caſe of Mrs. Hutchinſon. Two of her ſons refuſing to agree to her cenſure, were admoniſhed, and the church was thereby enabled to proceed ὁμοθυμαδὸν. — See *Winthrop*, i. 255.]

Hooker (*Survey*, iii. 40) lays down the rule that cenſure may be paſſed "*if ſome few ſhould diſſent*, in caſe their reaſons be heard and anſwered, and they ſilenced by power of argument;" and that, in doubtful caſes, if "the difference grow wide and great," after counſel of the neighbouring churches has been had, "either all will agree, or elſe *the major part of the church hath power and right to proceed*." Of his own prudent management under this rule, by which "he rarely miſſed of a full concurrence," and of its reſults, ſee the *Magnalia*, b. iii. pt. 1, app. § 25.

There is a touch of pathos in Cotton Mather's alluſion to the trials to which the "ſpeaking ariſtocracy" was occaſionally ſubjected, by the "ſilent democracy" of the congregation :

<small>ory is, and may be conftituted in England.</small> heare and end, or to certifie the Church, and any brethren, that will, to be prefent at the difcuffe in private.³⁸

<small>Difference in number of Officers.</small> Some Churches have no ruling Elders, fome but one, fome but one teaching Elder, fome have two ruling, and two teaching Elders; fome one, fome two or three Deacons; fome hold that one Minifter is enough for a fmall number of people; No Church there hath a Deaconeffe, as far as I know.³⁹

<small>Chappels of eafe.</small> Where farmes or villages are, as at *Rumney-marfh*⁴⁰ and *Marblehead*,⁴¹ there a Minifter, or a brother of one of the

---

"Now tho' this liberty of the brethren [to judge in their own church cafes], be that wherein for the moft part the *repofe* of the paftors has been by the compaffionate wifdom of our Lord Jefus Chrift provided for, yet fome *trouble* fometimes has arifen to the paftors from the brethren's abufe of their liberty, *which has call'd for much patience* in thofe that have the rule over them." — *Magnalia*, b. iv. pt. 2. c. iv. § 10.

³⁸ Comp. Cotton, *Way*, 95, 96.

³⁹ See before, p. 8, note 14.

⁴⁰ Now Chelfea. No church was gathered there until 1715. Early in 1640, the owners of farms at Rumney Marfh made requeft to the Bofton church, that John Oliver, (fon of Elder Thomas,) "a gracious young man," might be fent "to inftruct [their] fervants, and be a help to them, becaufe they cannot many times come hither,

nor fometimes to Lynn, and fometimes nowhere at all." The confent of the church was given, after fome debate, and Sergeant Oliver expreffed his willingnefs to "employ his weak talent to God's fervice." Savage, from *Keayne's Ms.*, in note to *Winthrop*, i. 328. Mr. Oliver died in 1646, — "one who, for the fweetnefs of his difpofition and ufefulnefs, through a public fpirit, was generally beloved, and greatly lamented." — *Ibid.* ii. 257.

⁴¹ "Marvill Head is a place which lieth four miles full fouth from Salem, and is a very convenient place for a plantation, efpecially for fuch as will fet upon the trade of fifhing. There was made here a fhip's loading of fifh the laft year, where ftill ftand the ftages and drying fcaffolds. Here be good harbour for boats, and fafe riding for fhips." Wood, *N. E. Prof-*

congregations of *Bofton* for the *Marſh*, and of *Salem* for
Marblehead, preacheth and exerciſeth prayer every Lords
day, which is called propheſying in ſuch a place. And ſo
it was heretofore at *Mountwoollaſton* within *Bofton* pre-
cinéts, though ſince it became a Church now called of
Braintree,[42] but before they of the mount did, and thoſe
of the *Marſh* and *Marblehead* ſtill come and receive the
Sacrament at *Boſton*, and *Salem* reſpectively, and ſome of
Braintree ſtill receive at *Boſton*.

Alſo when a Miniſter preacheth abroad, in another
congregation, the ruling Elder of the place, after the
Pſalme ſung, ſaying publiquely; If this preſent brother
hath any word of exhortation for the people, at this time,
in the name of God let him ſay on;[43] this is held proph-

Theſe, you ſee, are neceſſary in England in ſome places.

Propheſying.

Propheſying, or Preaching by Licence.

---

pect, pt. i. c. 10. The plantation was ſet off from Salem, as a ſeparate town-ſhip, in 1649. Joſſelyn found there "a few ſcattered houſes . . . ſtages for fiſhermen, orchards and gardens; half a mile within land, good paſtures and arable land." — *Voyages*, 167.

[42] The inhabitants of Mount Wol-laſton were granted town privileges, May, 1640, and the name of Brain-tree given. — *Maſs. Rec.*, i. 291. The church was gathered September 17, 1639, when Mr. William Tompſon and Mr. Henry Flint were choſen their miniſters. The former was or-dained November 19; Mr. Flint not until March 17, 1640. He had been

one of the ſigners of the remonſtrance againſt Wheelwright's cenſure, but "acknowledged his failing, and de-ſired his name might be blotted out," May, 1640. Mr. Savage ſuggeſts that his ordination at Braintree may have been poſtponed "to afford him liberal opportunity for this recantation." It is poſſible that his ſin of charity, though repented of, may have left a taint of error which influenced "ſome of Braintree" to receive the ſacra-ment at Boſton, after the gathering of a church in their own town. — *Maſs. Rec.*, i. 191; *Winthrop*, i. 196, 247, 313, 324.

[43] "The elders calling to them . . .

Plaine dealing.

<small>It ought not to be otherwaies in England.</small>

16

<small>* Univerſities, Cathedrals, and Collegiat Churches.</small>

<small>' 1 Cor. 13. 2.</small>

efying.⁴⁴ Alſo the confeſſions or ſpeeches made by members to be admitted, have beene by ſome held prophefying, and when a brother exerciſeth in his | own congregation (as at *Salem*⁴⁵ they doe ſometimes) taking a text of Scripture, and handling the ſame according to his ability. Notwithſtanding, it is generally held in the *Bay*, by ſome of the moſt grave and learned men amongſt them, that none ſhould undertake to prophefie in publique, unleſſe he intend the worke of the Miniſtery, and ſo in ſome places, as in ſchooles*, and not abroad, without they have both impoſition of hands, and miſſion, or permiſſion, becauſe prophecie properly hath its denomination from *underſtanding propheticall Scriptures*, which to know diſcreetly to handle, requireth good learning, ſkill in tongues, great fidelity, and good conſcience.⁴⁶

If they have any word of exhortation to the people, to ſay on." — Cotton, *True Conſtit. of a Particular Viſible Church*, p. 6.

⁴⁴ As to the diſtinction between "teaching by office" and "prophefying," ſee Ainſworth's *Counterpoyſon*, 1608, pp. 174–178; Robinſon's *Apology*, 1625, c. viii.; and *People's Plea for the Exerciſe of Prophcſy*, 1618, pp. 6, 33; Cotton, *Keyes*, 20, 21 (comp. Goodwin and Nye, in *Preface*); [or, in Hanbury's *Memorials*, i. 175–6, 353, 389; ii. 263;] Bradford's *Dialogue*, in Young's *Chron. of the Pilgrims*, 419, 420.

⁴⁵ " Mr. Skelton, the paſtor of Salem, and Mr. [Roger] Williams, who was removed from Plimouth thither, (but not in any office, though he exerciſed by way of prophecy,)" etc. — *Winthrop*, i. 117 (1633).

⁴⁶ "Though wee deny not, but in ſome caſe, ſome able judicious experienced Chriſtians, may humbly and ſoberly, when neceſſity requires, as in the want of Miniſters and being invited thereunto, diſpence now and then a word of exhortation to their brethren, This is farre enough from Preaching in an ordinary way [or, as W. R. had aſſerted,] *with all Au-*

The publique worſhipe.

THE publique worſhip is in as faire a *meeting houſe* as they can provide, wherein, in moſt places, they have beene at great charges.[47] Every Sabbath or Lords

thority."—Welde's *Anſwer to W. R.*, 37, 38.

Mr. Cotton accorded a larger liberty: "As for the publike teaching of a private man, indued with gifts and zeal, I know not why it may not be allowed, not only in caſe of extreme neceſſitie, but in ſome caſes of expediency, as when his gifts are to be proved before he be called into office." (*Way cleared*, ii. 24.) "It is not an unheard of novelty, That God ſhould enlarge private men with publike gifts, and that they that have received ſuch gifts, ſhould take liberty to diſpenſe them unto edification." (*Ibid.* 27.) "And in this," ſays Gov. Bradford, "the chief of our miniſters in New England agree."—*Dialogue*, &c., in Young's *Chron.* 421.

When Mr. Wilſon went to England, in 1631, he commended to his church, "the exerciſe of propheſy in his abſence, and deſigned thoſe whom he thought moſt fit for it," namely, Gov. Winthrop, Mr. Dudley, and the ruling elder, Increaſe Nowell. —*Winthrop*, i. 50.

The next year, when Winthrop was in Plymouth on the Sabbath, Mr.

Roger Williams propounded a queſtion, "according to their cuſtom," "to which the paſtor, Mr. Smith, ſpake briefly; then Mr. Williams propheſied," and, afterwards, Gov. Bradford, Elder Brewſter, "then ſome two or three more of the congregation," and, by invitation, Gov. Winthrop and Mr. Wilſon, ſpoke to the queſtion.—*Winthrop*, i. 91, 92.

In 1634, when the people at Agawam (Ipſwich) were without a miniſter, Gov. Winthrop "ſpent the Sabbath with them, and exerciſed by way of prophecy."—*Ibid.* i. 30.

"The practice of private members making ſpeeches in the church aſſemblies, to the diſturbance and hindrance of the ordinances," was one of the evils reproved by Mr. Rogers of Rowley, in his ſermon before the Synod and the General Court, in 1647.—*Winthrop*, ii. 308.

47 The new meeting-houſe in Boſton was finiſhed the year before Lechford's departure. It ſtood (for ſeventy-one years) "on the ſite now occupied by Joy's Building, in Waſhington Street, a little to the ſouth of, and oppoſite to, the head of State Street."

Every Sunday morning. day, they come together at *Boston*, by wringing of a bell,[48] about nine of the clock or before. The Pastor begins

Drake's *Boston*, 142. It was erected at a cost of about £1000, "which was raised out of the weekly voluntary contribution, without any noise or complaint." (*Winthrop*, ii. 24.) Joshua Scottow, contrasted the "amplified and dignified" church of Boston, in his latter days, with "that little church which after seven years growth, its number *(in their mud-wall Meeting-House, with wooden Chalices)* was so small as a child might have told [counted] the whole Assembly." — *Narr. of the Planting*, &c. (4 *Mass. Hist. Coll.*, iv. 307).

[48] Boston was favored, in having a bell "to wring," in 1641, or before,— though Lechford does not tell us whether the bell was *stationary*, or *perambulatory* in the hand of a bell-man. In most of the towns of New England, at this period, the summons to public worship, and to other meetings of the inhabitants, was given by beat of drum. Johnson relates, how a new-comer from England, in 1636, when near Cambridge, "hearing the sound of a Drum, . . . demands of the next man he met what the signall of the drum ment; the reply was made that they had as yet no Bell to call men to meeting; and thereupon made use of a drum." *W. W. Providence,* b. i. c. xliii. Yet Prince states, on the authority of a manuscript letter, that the Cambridge meeting-house,

built in 1632, had "a bell *upon it;*" and Dr. Holmes thinks the statement confirmed by the town-records, which show that town-meetings were then called by the ringing of the bell. *Hist. of Cambridge; Mass. Histor. Coll.*, vii. 19. Mr. Davenport of New Haven, writing to Gov. Winthrop, October 17, 1662, mentions the sickness of his colleague, Mr. Street, who, "the last lecture day . . . purposed to preach . . . and continued in that purpose till the *second drum*, but then was compelled to take his bed." Another letter (November, 1660) gives an account of the last sickness of Gov. Newman : "My son went to him after the beating of the first drum. . . . When the second drum beat, I was sent for to him."

Hartford had a town-crier and bell-ringer as early as 1641, at least; and in 1643, the town ordered "a bell to be rung by the watch every morning, an hour before day break," and "that there should be in every house, one up, and have made some light, within one quarter of an hour after the end of the bell ringing." To devise a penalty that would insure compliance with such a requisition, in this generation, might prove a difficult problem for legislators. That Watertown had a church-bell as early as February, 1649, the payment at that time for a bell-rope, which is noted in the town

with folemn prayer continuing about a quarter of an houre. The Teacher then readeth and expoundeth a Chapter;[49] Then a Pfalme is fung, which ever one of the ruling Elders dictates.[50] After that the Paftor preacheth a Sermon,[51] and fometimes *ex tempore* exhorts. Then the Teacher concludes with prayer, and a bleffing. Once a moneth is a Sacrament of the Lords Supper,[52] Lords Supper.

records, feems to prove. (Bond's *Watertown*, 1046.)

[49] " After prayer, either the paftor or teacher readeth a chapter in the Bible, and expoundeth it."— Cotton, *Way*, 67. Comp. *True Conflit. of a Church*, 6.

" In England," wrote Lechford to a friend in 1640, " twelve or thirteen chapters and pfalms are read every Sunday, in all churches, befide what is upon Wednefdays and Fridays and other holydays; but here, Scripture twice a Sunday, in any Church, upon whatfoever occafion ; but preaching, and long conceived prayers."— *Ms. copy (in fhort-hand)*. Comp. p. 20, after.

[50] " Before Sermon, and many times, after, we fing a Pfalme, and becaufe the former tranflation of the Pfalmes doth in many things vary from the original, and many times paraphrafeth rather than tranflateth ; befides divers other defects (which we cover in filence) we have endeavoured a new tranflation of the Pfalmes into Englifh meetre, as near the originall as wee could exprefs it, ... and thofe Pfalmes we fing both in our publick churches, and in private."— Cotton, *Way*, 67.

[51] " In difpenfing whereof, the Minifter was wont to ftand above all the people in a pulpit of wood, and the Elders on both fides." Cotton, *True Conflit. of a Church*, 6. " In fundry churches, the other, whether paftor or teacher, who expoundeth not, he preacheth the Word ; and in the afternoon, the other who preached in the morning, doth ufually (if there be time) reade and preach, and he that expounded in the morning preacheth after him."— *Way*, 67.

" At Quinnipyack [New Haven] Mr. Davenport preached in the forenoon that men fhould be uncovered, and ftand up at the reading the text ; and in the afternoon the affembly jointly practifed it."— Mr. Hooker, in letter to Mr. Shepard, March 20, 1640 [in *Hutchinfon*, i. 430, note].

[52] Comp. Cotton, *Way*, 67-69.

17 whereof notice is given usually a fortnight | before, and then all others departing save the Church, which is a great deale lesse in number then those that goe away, they receive the Sacrament, the Ministers and ruling Elders sitting at the Table, the rest in their seats, or upon forms: All cannot see the Minister consecrating, unlesse they stand up, and make a narrow shift. The one of the teaching Elders prayes before, and blesseth, and consecrates the Bread and Wine, according to the words of Institution; the other prays after the receiving of all the members: and next Communion, they change turnes; he that began at that, ends at this: and the Ministers deliver the Bread in a Charger to some of the chiefe, and peradventure gives to a few the Bread into their hands, and they deliver the Charger from one to another, till all have eaten; in like manner the cup, till all have dranke, goes from one to another. Then a Psalme is sung, and with a short blessing the congregation is dismissed. Any one, though not of the Church, may, in *Boston*, come in, and *see the Sacrament administered, if he will:[53] But none of any Church in the Country may receive the Sacrament there, without leave of the congregation, for which purpose he comes

* Once I stood without one of the doores, and looked in, and saw the administration: Besides, I have had credible relation of all the particulars from some of the members.

[53] " It is not true that wee hold out any at all, English or Indian, out of our Christian Congregations. All without exception are allowed to be present, at our publick Prayers and Psalmes, at our reading of the Scrip-

to one of the ruling Elders, who propounds his name to the congregation, before they goe to the Sacrament.[54] About two in the after-noone, they repaire to the meet- ing-houfe againe; and then the Paftor begins, as before noone, and a Pfalme being | fung, the Teacher makes a Sermon. He was wont, when I came firft, to reade and expound a Chapter alfo before his Sermon in the afternoon. After and before his Sermon, he prayeth. After that enfues Baptifme,[55] if there be any, which is

Afternoone.

18

Baptifme.

tures, and the preaching and expounding of the fame, and alfo at the admitting of Members and difpenfing of feales and cenfures." — Cotton, *Way cleared*, i. 69.

[54] "The members of any Church, if any be prefent, who bring Letters teftimoniall with them to our Churches, wee admit them *to the Lords Table* with us, and their *children* alfo (if occafionally in their travell they be borne with us) upon like recommendation, wee admit to *Baptifme*." Cotton's *Way of the Churches*, 68. Compare, *Keyes*, 17; Hooker's *Survey*, iii. 28, 29, 32; *Anfwer to Nine Pofitions*, 17; *Defence of the Anfwer*, by Allin and Shepard, ch. iii. 2. "We hold it not unlawfull, (but doe often practife) to receive other members to communion with us *without letters;* efpecially if they bee knowne to any of our Church, elfe fuch letters are defirable." — Welde's *Anfwer to W. R.*, 53.

[55] Compare Cotton's *Way of the Churches*, 67, 68. Hooker (*Survey*, iii. 28) fays that the Lord's Supper and Baptifm "muft be difpenfed publikely, in the prefence, and with the concurrence of the Church folemnly affembled," and fhould "goe hand in hand" with preaching; "after the word preached, the feals fhould be adminiftred." So, the New Haven Church Catechifm, by Davenport and Hooke (repr. New Haven, 1853, p. 56), in anfwer to the queftion, "How is Baptifm to be adminiftered?" I do not find, in the early authorities on Congregational order, an intimation that baptifm might not rightfully be adminiftered on any day of the week, when the Church was affembled and the word preached. See *Anfwer to Nine Pofitions*, pp. 36, 37. Mr. Ball, in the *Reply to the Anfwer* (p. 38), remarks incidentally, and not as if the pofition was a matter of controverfy, "Baptifme is not tyed to the

done, by either Paftor or Teacher, in the Deacons feate, the moft eminent place in the Church, next under the Elders feate. The Paftor moft commonly makes a fpeech or exhortation to the Church, and parents concerning Baptifme, and then prayeth before and after. It is done by wafhing or fprinkling. One of the parents being of the Church, the childe may be baptized, and the Baptifme is into the name of the *Father*, and of the *Sonne*, and of the *holy Ghoft*. No fureties are required.

Contribution. Which ended, follows the contribution, one of the Deacons faying, Brethren of the congregation, now there is time left for contribution, wherefore as God hath profpered you, fo freely offer.[56] Upon fome extraordinary

firft day of the week." That, in point of fact, this facrament was *ufually* — perhaps, *almoft invariably* — adminiftered on the firft day, in the churches of New England, there is no room for doubt. Mr. Davenport, writing, in 1666, about the innovations which the Rev. Jofeph Haynes was introducing in the church at Hartford, fays, parenthetically, that he fuppofes baptifm "was never adminiftered, in a week day, *in that Church*, before." 3 *Mafs. Hift. Coll.*, x. 61. But the "lax ways" which (in the fame letter) he cenfured in Mr. Haynes were thofe which concerned the *fubjects* of baptifm, not merely the *time* of its adminiftration. Mr. Davenport was a zealous "Anti-synodift," or oppofer of the Half-way Covenant. Mr. Haynes went even *beyond* the Synodifts in "large Congregationalifm" (as it was afterwards termed), by admitting not only the children of half-way covenanting parents, but grandchildren in right of covenanting grandparents, adopted children, fervants; and flaves, in right of their adoptants and mafters.

[56] "The *Deacons*, (who fit in a feate under the *Elders*, yet in fundry churches lifted up higher then the other pewes,) doe call upon the people, that as God hath profpered them, and hath made their hearts willing, there is now time left for contribution."—Cotton, *Way of the Churches*, 69.

occafions, as building and repairing of Churches or meeting-houfes, or other neceffities, the Minifters preffe a liberall contribution, with effectuall exhortations out of Scripture. The Magiftrates and chiefe Gentlemen firft, and then the Elders, and all the congregation of men and moft of them that are not of the Church, all fingle perfons, widows, and women in abfence of their husbands, come up one after another one way, and bring their offerings to the Deacon at his feate, and put it into a box of wood for the purpofe, if it bee money or | papers ; if it be any other chattle, they fet it or lay it downe before the Deacons, and fo paffe another way to their feats againe.[57] This contribution is of money, or papers, promifing fo much money: I have feene a faire gilt cup with a cover, offered there by one, which is ftill ufed at the Communion. Which moneys, and goods the Deacons difpofe towards the maintenance of the Minifters, and the poore of the Church, and the Churches occafions, without making account, ordinarily.[58]

[57] "The people from the higheft to the loweft in fundry Churches do arife, the firft pew firft, the next next, and fo the reft in order, and prefent before the Lord their holy offerings." *Ibid.* Comp. Joffelyn, *Voyages*, 180. In Brewfter's church at Plymouth, when Gov. Winthrop was there in 1632, "The deacon, Mr. Fuller, put the congregation in mind of their duty of contribution ; whereupon the governour and all the reft went down to the deacon's feat, and put into the box, and then returned."—*Winthrop*, i. 92.

[58] "This weekly contribution is properly intended for the poore, according to 1 *Cor.* 16. 1. Yet fo as (if there be much given in,) fome churches doe (though others do not) appoint

Differences in contributions.

But in *Salem* Church, thofe onely that are of the Church, offer in publique; the reft are required to give to the Minifterie, by collection, at their houfes. At fome other places they make a rate upon every man, as well within, as not of the Church, refiding with them, towards the Churches occafions; and others are beholding, now and then, to the generall Court, to ftudy wayes to enforce the maintenance of the Minifterie.[59]

the overplus *towards* the Minifters maintenance. 2. This is not given in by the people *according to their weekly gaines* [as Rathband had ftated,] but *as God hath bleft them with an eftate in the generall.* ... 3. Nor is this difpenfed to the Minifters (in thofe churches where any part of it is fo given) though by the hands of the Deacons, yet not for proportion as they pleafe, ... but by the Church, who ufually, twice in the year or oftener, doe meete to confult and determine of the fumme to be allowed for that yeere to their Minifters, and to raife it, either from the Churches treafurie . . . . or by a contribution to be then made on purpofe."—Welde, *Anfwer to W.R.*, 59.

[59] See the order of court, Sept., 1638, *Maff. Rec.*, i. 240. Mr. Cotton was not willing that the Bofton Church fhould avail itfelf of any compulfory procefs, and taught his people, "that when magiftrates are forced to provide for the maintenance of minifters, etc., then the churches are in a declining condition;" and "he fhowed that the minifters' maintenance fhould be by voluntary contribution," &c. *Winthrop*, i. 295. When Roger Williams objected to the "conftraint laid upon all confciences . . . to come to church and pay church duties," (*Bloudy Tenent*, c. lxix.) Mr. Cotton replied, "I know of no conftraint at all that lieth upon the confciences of any in New England, to come to Church . . . Leaft of all do I know that any are conftrained to pay church duties in New England. Sure I am none in our own town are conftrained to pay any church duties at all. What they pay they give voluntarily, each one with his own hand, without any conftraint at all but their own will, as the Lord directs them." (*Bl. Tenent Wafhed*, 146.) In his rejoinder, Williams fays: "For a freedom of not paying in his [Mr. Cotton's] town, *it is to their commendation, and God's praife.* Yet who can be ignorant of the affeffments upon all, in *other* towns," etc. (*Bl. Tenent yet*

This done, then followes admiffion of members, or hear- [Admiffions / Offences]
ing matters of offence, or other things, fometimes [60] till it
be very late. If they have time, after this, is fung a
Pfalme, and then the Paftor concludeth with a Prayer and
a bleffing.

Upon the week dayes, there are Lectures in divers [Lectures / Fafts & Fealts]
townes,[61] and in *Bofton*, upon Thurfdayes, when Mafter

more bloody, 216.) It is not eafy to reconcile Mr. Cotton's general denial with Winthrop's ftatement, (ii. 93,) that fome churches raifed their minifters' maintenance by taxation, "which was very offenfive to fome;" or with his account of the profecution of "one Brifcoe of Watertown, who ... being grieved ... becaufe himfelf and others, who were no members, were taxed, wrote a book againft it," which he "publifhed underhand;" for which offence the court fined him £10, and "one of the publifhers" £2, in March, 1643,—not long before Roger Williams failed for England (where he printed the Bloudy Tenent).

Hooker, (*Survey*, ii. 29, 32,) regarding it the duty of "every one that is taught" to contribute, argues that fuch contribution fhould be enforced, not by the civil magiftrate, but by the difcipline of the church. "In cafe any member fhall fail in this free contribution, he finnes in a breach of the knowne rule of the Gofpell; it appertains to the Church, to fee the Reformation of that evill, as of any other fcandall." And he makes it the duty of the deacon, if any member fail to perform this duty, to admonifh, and, in cafe he reform not, to "follow the action againft him ... and bring him to the cenfure of the church." *Ibid.* 37.

[60] In the M.H.S. MS., the comma is placed *after* 'fometimes;' "or other things fometimes, till," &c.

[61] "So that fuch whofe hearts God maketh willing, and his hand doth not detaine by bodily infirmitie, or other neceffary imployments, (if they dwell in the heart of the Bay) may have opportunitie to heare the Word almoft every day of the weeke in one Church or other, not farre diftant from them." Cotton, *Way of the Churches*, 70. In 1639, "there were fo many lectures ... and many poor perfons would ufually refort to two or three in the week, to the great neglect of their affairs and the damage of the public,"—that the General Court fought a conference with the Elders "to confider about the length and frequency of church affemblies," &c.

Cotton teacheth out of the *Revelation*.[62] There are dayes
of fasting, thanksgiving, | and prayers upon [a]occasions, but
no [b] holy dayes, except the Sunday.

[a] And why not set failing dayes & times, and set feasts, as well as set Synods in the Reformed Churches? [b] And why not holy dayes as well as the fift of November, and the dayes of Purim among the Jews? Besides, the commemoration of the blessed and heavenly mysteries of our ever blessed Saviour, and the good examples and piety of the Saints? What time is there for the moderate recreation of youth and servants, but after divine services on most of thofe dayes, seeing that upon the Sunday it is juftly held unlawfull? And sure enough, at New-England, the Masters will and muft hold their servants to their labour more then in other Countries well planted is needfull; therefore I think even they should doe well to admit of some Holy dayes too, as not a few of the wifer fort among them hold neceffary and expedient.

But "this was taken in ill part by most of the elders and others of the churches," who regarded it as an infringement of their liberties, and feared it might "also raise an ill favour of the people's coldnefs, that would complain of much preaching," — and the magiftrates "finding how hardly fuch propofitions would be digefted . . . thought it not fit to enter any difpute or conference on the fubject." (*Winthrop*, i. 324, 325.) Rarely, fince then, has the General Court had occafion to confider the expediency of legiflating for the fuppreffion of inordinate churchgoing.

[62] Mr. Cotton's fermons upon the thirteenth chapter of the Revelation were printed in London, in 1655, from notes taken by one of his hearers. An Epiftle to the Reader, by Rev. Thomas Allen (formerly of Charleftown, but then of Norwich, co. Lincoln), fpeaks of his having had "the happy priviledg [while living in that American wilderneffe . . . in the towne next adjoyning to *Bofton*,] of enjoying the benefit of the precious labours of Mr. *Cotton*, in his Lecture upon every fifth day of the week;" and ftates that this expofition of chap. xiii. was delivered "about the 11. and 12. moneths (if I miftake not) of the year 1639, and the firft and fecond of the yeare 1640." Before June, 1641, Mr. Cotton had reached the end of the 15th chapter. (*Winthrop*, ii. 30.) His Sermons on the Seven Vials, from the 16th chapter, were printed early in 1642, and the volume was received in Bofton in July. (*Ibid*., ii. 75.) "Mr. Humfrey had gotten the notes from fome *who had took them by characters*, and printed them in London," without Mr. Cotton's confent. Was this note-taker Lechford?—"This Venerable Seer," wrote Jofhua Scottow, "*whofe method was to go through the Books of Scripture he entred upon*, and had in his Minifterial Courfe in both *Boftons* been (lengthened out to little lefs than forty years), went through near the whole Bible." —*Narrative of the Planting*, &c., 4 *Mafs. Hift. Coll*., iv. 284.

Newes from New-England. 53

In some Churches, nothing is ᶜread on the first day of the weeke, or Lords day, but a Psalme dictated before or after the Sermon, as at *Hing-ham;* there is no catechizing of children or others in any Church, (except in *Concord* Church, & in other places, of those admitted, in their receiving:) the reason given by some is, because when people come to be admitted, the Church hath tryall of their knowledge, faith, and repentance, and they want a direct Scripture for Ministers catechizing;⁶⁵ as if, *Goe teach all Nations,* and *Traine up a childe in the way he should goe,* did not reach to Ministers catechizings. But, God be thanked, the generall Court was so wise, in *Iune* last, as to enjoyn, or take

[Little reading,⁶⁷ catechizing.

ᶜ Whereas in *England* every Sunday are read in publique, Chapters and Psalmes in every Church, besides the eleven or twelve Commandements,⁶⁴ Epistle and Gospell, the Creed and other good formes and catechizings, and besides what is read upon Holy dayes and other dayes both in the parish, and Cathedrall and Collegiate Churches, & in the Universities, and other Chappels, the benefit whereof, doubtlesse, all wise men will acknowledge to be exceeding great, as well as publique preaching and expounding.]

⁶³ See before, p. 16, note 49.
⁶⁴ The printer of the first edition misplaced the words "eleven or twelve," which should have been inserted, in the line above, before "Chapters." The manuscript in the Library of the Massachusetts Historical Society is sufficient authority, if any is required, for the correction of so obvious an error. In all the copies of the first edition which I have seen (as in that from which the Mass. Hist. Society's reprint was made), "a pen has been carefully drawn through the words 'eleven or twelve,' and the color of the ink shows this to have been done early." (3 *Mass. Hist. Coll.,* iii. 79.)—For "Creed," in the third line of the note, the M.H.S. MS. has "creeds;" and, in the fifth line, after "parish," is the word "churches."

⁶⁵ "The excellent and necessary use of catechising young men, and novices, . . . we willingly acknowledge: But little benefit have wee seene reaped from set formes of questions, and answers devised by one Church, and imposed by necessity on another."—Cotton, *Ans. to Ball's Disc. of Set Formes of Prayer,*(London, 1642,) ch. vii. p. 41.

some course for such catechizing, as I am informed, but know not the way laid down in particular, how it should be done.[66]

21
Dayes and moneths how called.

They call the dayes of the weeke, beginning at the first, second, third, forth, fifth, sixth, and seventh, which is Saturday: the moneths begin[67] at March, by the names of the first, second, and so forth to the twelfth, which is *February*: because they would *avoid all memory* of heathenish and idols *names*: And surely it is good to overthrow heathenisme by all good[68] wayes and meanes. But there

Neglect of instructing the Indians.

hath not been any sent forth by any Church to learne the Natives language, or to instruct them in the Religion;[69]

---

[66] "It is desired that the elders would make a catechisme for the instruction of youth in the grounds of religion." (June, 1641.) — *Mass. Rec.*, i. 328.

[67] "Beginning." — *M.H.S. MS.*

[68] The insertion of the word "good" — which is not in the M.H.S. MS. — was a judicious qualification.

[69] A few years later, the labors of Eliot, Mayhew, and their fellow-laborers, had done much to remove this reproach from the churches. Robert Baylie, the Presbyterian writer, in "*A Dissuasive from the Errours of the Time*," printed in 1645, cites this remark of Lechford's in support of his averment that the Independents of New England, "of all that ever crossed the *American* Seas, . . . are noted as most neglectful of the work of Conversion" (p. 60). Mr. Cotton, in *The Way of Congregational Churches cleared* (1648), — which was written in reply to Baylie's book, — asks, "What if there have not bin any sent forth by any Church to learn the *Indians* language? That will not argue our neglect of minding the work of their conversion. For there be of the *Indians* that live amongst us, and dayly refort to us; and some of them learne our language; and some of us learn theirs. And men that love the Lord Jesus doe gladly take opportunity to instruct them in our Religion, and to teach them both Law and Gospell. And of late, the Word (as I have said) is publickly preached unto them in two severall

First, becaufe they fay they have not to do with them being without, unleffe they come to heare and learn Englifh.[70] Secondly, fome fay out of *Rev.* 15. laft,[71] it is not probable that any nation more can be converted, til the calling of the Jews; *till the feven plagues finifhed none was able to enter into the Temple*, that is, the Chriftian Church, and the *feventh Viall* is not yet poured forth, and God knowes when it will bee.[72] Thirdly, becaufe all Churches among them are equall, and all Officers equall, and fo betweene many, nothing is done that way. They muft all

Indian Congregations [one neer to *Dorchefter Mill*, and another in Cambridge, neer *Watertown Mill*], though wee never thought it fit to fend any of our Englifh to live amongft them, to learn their language: *for who fhould teach them?*"—Pt. i. pp. 78, 79.

To High-Church writers, as well as to Prefbyterians, Lechford's ftatements on this point fupplied convenient matter of reproach againft New England. "O that we could approve to God and our confciences that [the propagation of Chriftian religion] is our main motive and principal drift in our Weftern plantations; but how little appearance there is of this holy care and endeavour, the plaine dealer upon knowledge hath fufficiently informed us."—*Diverfe Practicall Cafes of Confcience Refolved* (1649), p. 323, cited by Dr. Palfrey, *Hift. of N. England*, ii. 192. (See after, p. 69, and note.)

[70] "I know not whether ever any gave him fo weake an account, or no: If any fo did, it was his rafhneffe, or ignorance both of us, and the truth. But if the Author fpeake it, as a Point of our Profeffion or practife, that we doe neglect the inftruction of the *Indians*, and efpecially upon fuch a reafonleffe reafon, I will fay no more to it but this, it feemeth there are two forts of *Plain dealing*: Plain honeft dealing, and Plain falfe dealing, of which latter fort, this fpeach is."— Cotton's *Way cleared*, pt. i. p. 79.

[71] "Verfe the laft."—*M.H.S. MS*.

[72] This was Mr. Cotton's belief. See *Winthrop*, ii. 30; (and comp. *Way of the Churches cleared*, pt. i. p. 78.) "Till the Jewes come in, there is a feale fet upon the hearts of thofe people, as they thinke from fome Apocalypticall places." *The Day-Breaking*, &c., pp. 15, 16. — Roger Williams, though (at this period) he

Plaine dealing,

therefore equally beare the blame; for indeede I humbly[73] conceive that by their principles, no Nation can or could ever be converted. Therefore, if so, by their principles how can any Nation be governed? They have nothing to excuse themselves in this point of not labouring with the Indians to instruct them, but their want of[74] a staple trade, and other businesses taking them up. And it is true, this may excuse *à tanto*.

22 Charity.

Of late some Churches are of opinion, that any | may be admitted to Church-fellowship, that are not extremely ignorant or scandalous: but this they are not very forward to practice, except at *Newberry*.[75] Besides, many good people scruple their Church Covenant, so highly

was more hopeful of the fruits of labors among the Indians, likewise believed "that no remarkable conversion of the nations is yet to be expected, because smoke filled the temple till Antichrist was overthrown. *Rev.* xv. 8."—*Hireling Ministry none of Christ's*, p. 13; in Knowles's *Memoir of R. Williams*, 378.

[73] "I humbly" was substituted, on revision, for "some."—M.H.S. MS.

[74] "Their poverty, and want," &c. —*Ibid.*

[75] Where the Rev. James Noyes was pastor, and the Rev. Thomas Parker, teacher. Mr. Noyes "was jealous (if not too jealous) of particular Church-covenants. . . . He held . . . that such as show a willingness to repent, and be baptised in the name of the Lord Jesus, without known dissimulation, are to be admitted [to Church fellowship] . . . and that God took into covenant some that were vessels of wrath, as for other ends, so to facilitate the conversion of their elect children."—*Rev. Nich. Noyes*, in the *Magnalia*, b. iii. pt. 2. ch. 25. Mr. Parker wrote to a member of the Westminster Assembly, in December, 1643, that he and his colleague held that "the rule must be so large that the weakest Christians may be received; and [that] there was according to appearance, much conjunction in this particular," among those present in the Synod at Cambridge in September.—Hanbury's *Memorials*, ii. 295.

Newes from New-England. 57

tearmed by the moſt of them,[76] a part of the *Covenant of grace; and particularly, one Maſter *Martin*[77] for ſaying in argumentation, that their Church Covenant was an humane invention, and that they will not leave till it came to the fwords point, was fined ten pounds, his cow taken and fold for the money. A Miniſter[78] ſtanding upon his

*The Covenant of Grace of the New Teſtament, it is true, makes the whole univerſall Church of Chriſt, and every part thereof, or at leaſt belongeth thereunto: but allowing Churches a Covenant of

Mr. Warham, of Dorcheſter, and afterwards of Windſor, Conn., appears to have favored "the pariſh way" (as it was called) at his firſt coming to New England. Samuel Fuller wrote to Bradford, June, 1630, that "Mr. Warham holds that the viſible church may confiſt of a mixed people,—godly, and openly ungodly; upon which point we all had our conference," &c. — 1 *Maſs. Hiſt. Coll.*, iii. 74.

[76] The earlier (M.H.S.) MS. has "by them," without the words "the moſt of." — According to Hooker, the church-covenant is "an *ordinance* of the Goſpel, and *warranted* by the Goſpel, but it is *not* in propriety of ſpeech *the covenant of the Goſpel*." "A man may be in the covenant of grace, and ſhare in the benefit thereof, who is not in a Church ſtate, and a man may be in a Church ſtate, who is not really in the Covenant of grace . . . and therefore the one is not the other." — *Survey*, i. 70, 78, 79.

[77] The M.H.S. MS. ends the paragraph abruptly with, "and Mr. Martin." The fourteen lines which follow were transferred to this place, on reviſion, with additions and alterations,

from another page (27) of that MS., where they follow the notice of the church at Aquedney, under the caption, "Mr. Lenthall his Controverſie. Brittaine whipt,"—and a marginal direction (in ſhort-hand), "Put theſe notes in their proper places."

At a General Court, March 13, 1639,—"Mr. Ambros Marten [of Dorcheſter,] for calling the church covenant a ſtinking carrion & a humane invention, & ſaying hee wondered at God's patience, feared it would end in the ſharpe, & ſaid the miniſters did dethrone Chriſt, & ſet up themſelves; hee was fined 10*l.* and counſelled to go to Mr. Mather, to bee inſtructed by him." — *Maſs. Rec.*, i. 252. See *Winthrop*, i. 289.

[78] Mr. Robert Lenthall.— See, after, p. 41, note 144; *Winthrop*, i. 287, 288; *Maſs. Rec.*, i. 217, 254. Mr. Lenthall did not long remain in the Colony. In 1640, he was at Newport; admitted a freeman there Aug. 6, and employed by the town to teach a public ſchool. He returned to England in 1641 or 1642.— Arnold's *Hiſt. of R. Iſland*, i. 145-46; Callender's *Hiſt. Diſcourſe*, 62. It

Ministery, as of the Church of England, and arguing against their Covenant, and beeing elected by some of *Weymouth* to be their Minister, was compelled to recant some words; one that made the election, & got hands to the paper, was fined 10. pounds,[79] and thereupon speaking a few crosse words, 5. pound more, and payed it downe presently; Another[80] of them for saying one of the Ministers of the Bay was a Brownist, or had a Brownisticall head, and for a supposed lie, was whipt: and all these by the generall or quarter civill Courts.

Reformation tending to the better ordering and well-being of themselves, and for other politique respects, this is as much as they at New-England can justly make of their Covenant, and some that are judicious among themselves have acknowledged it: And yet, even this, unlesse it be made and guided by good counsell, and held with dependance and concatenation upon some Chiefe Church or Churches, may tend to much division and confusion, as is obvious to the understanding of those that are but a little versed in study of these points.

23

Ecclesia regnans.

Touching the government of the Common-Weale there.

NOne may now be a *Freeman* of that Commonwealth, being a Societie or Corporation, named by the name of the *Governour, Deputy Governour, and Assist-*

was after his return, probably, that Lechford erased the note made in his earlier draught, — that at Aquedney, "is Mr. Lenthall, a minister out of office, and lives very poorly." — *M.H.S. MS.*

[79] John Smith, "a chief stirrer in the businefs," was fined £20, at the March court, 1639. The fine not being paid, the May court fined him £5 for contempt, and ordered him to be imprisoned till both fines should be paid; but "on his submission, and bringing in of his money," the court remitted £10 of the amount, "and so only £15 was taken." — *Mass. Rec.*, i. 252, 254, 258; *Winthrop*, i. 288.

[80] "One of them named Brittaine." — *M.H.S. MS.* p. 27. James Britton, "who had spoken disrespectfully of the answer which was sent to Mr. Barnard his book against our church covenant, and of some of our elders, and had sided with Mr. Lenthall, etc., was openly whipped, becaufe he had no estate to answer, etc." — *Winthrop*, i. 289. The whipping did not produce a thorough reformation of manners, — for Britton was hanged for adultery, March 1, 1644. — *Ibid.*, ii. 158, 159.

Newes from New-England.

ants of the Societie of the Mattachufets Bay in New England, unleffe he be a Church member amongft them. None have voice in elections of Governours, Deputy, and Affiftants; none are to be Magiftrates, Officers, or Jurymen, grand or petite, but *Freemen.* The Minifters give their votes in all elections of Magiftrates. [81] Now the moft of the perfons at *New-England* are not admitted of their Church, and therefore are not *Freemen*, and when they come to be tryed there, be it for life or limb, name or eftate, or whatfoever, they muft bee tryed and judged too by thofe of the Church, who are in a fort their adverfaries: how equall that hath been, or may be, fome by experience doe know, others may judge.

The manner of the elections is this: At firft, the chiefe Governour and Magiftrates were chofen in *London*, by erection of hands, by all the *Free-men* of this *Society.* Since the tranfmitting of the Patent into *New-England*, the election is not by voices, nor erection of hands, but by papers,[82] thus:

Elections of the Governour & chiefe Magiftrates.

[81] The seven lines which follow (to the end of the paragraph) are not in the M.H.S. MS.

[82] After the transfer of the government to New England, election of governor and affiftants continued to be made "by erection of hands" until 1634, when Dudley was chofen in the place of Winthrop, "by papers." — *Winthrop*, i. 132. "This is the firft inftance of an election by ballot. It would have been hard for the freemen to nerve themfelves to the point of difplacing their old benefactor by the cuftomary "erection of hands." — Palfrey, *Hift. of N. England*, i. 375.

In September, 1635, the General Court ordered, "that, hereafter, the deputyes to be chofen for the Generall Courts fhalbe elected by papers, as

The generall Court-electory fitting, where are prefent in the Church, or meeting-houfe at | *Bofton*, the old Governour, Deputy, and all the Magiftrates, and two Deputies or Burgeffes for every towne, or at leaft one, all the *Freemen* are bidden to come in at one doore, and bring their votes in paper, for the new Governour, and deliver them downe upon the table, before the Court, and fo to paffe forth at another doore. Thofe that are abfent, fend their votes by proxies.[83] All being delivered in, the votes are counted, and according to the major part, the old Governour pronounceth, that fuch an one is chofen Governour for the yeare enfuing. Then the *Freemen*, in like manner, bring their votes for the Deputy Governour, who being alfo chofen, the Governour propoundeth the Affiftants one after the other. New Affiftants are, of late, put in nomination, by an order of general Court, beforehand to be confidered of:[84] If a *Freeman* give in a

---

the Governor is chofen."—*Mafs. Col. Records*, i. 157.

[83] In March, 1636, liberty was granted to fuch freemen as, for the fafety of their towns, fhould be detained at home on the day of election, "to fend their voices by proxy." The next year, it was made "free and lawfull for all freemen to fend their votes for elections by proxie the next Generall Courte in May, and fo for hereafter." — *Winthrop*, i. 185; *Mafs. Records*, i. 166, 188.

[84] *Mafs. Records*, i. 293. This order, giving the nomination of new affiftants to the freemen, was made at the General Court in May, 1640, after the election of Dudley. The year previous, the governor (Winthrop) and magiftrates had given offence to the freemen by nominating Emanuel Downing (Winthrop's brother-in-law) and two others for election as affiftants. A fufpicion "that the magiftrates intended to make themfelves ftronger, and the deputies weaker,

blanck, that rejects the man named; if the *Freeman* makes any mark with a pen upon the paper which he brings, that elects the man named; then the blancks and marked papers are numbred, and according to the major part of either, the man in nomination ſtands elected or rejected. And ſo for all the Aſſiſtants. And after every new election, which is, by their Patent, to be upon the laſt Wedneſday in every Eaſter Terme,[85] the new Governour and Officers are all new ſworn. The Governour and Aſſiſtants chooſe the Secretary. And all the Court conſiſting of Governour, Deputy, Aſſiſtants, and Deputies of towns, give their votes as well as the reſt; and the *Miniſters*, | and *Elders*, and all *Church-officers*, have their votes alſo in all theſe elections of chiefe Magiſtrates. Conſtables, and all other inferiour Officers, are ſworn in the generall, quarter, or other Courts, or before any Aſſiſtant.

Every *Free-man*, when he is admitted, takes a ſtrict oath, to be true to the Society, or juriſdiction: In which

25

Freemen their oath.

and ſo, in time, to bring all power into [their own] hands," occaſioned ſome oppoſition to Winthrop's re-election in 1639, and doubtleſs contributed to effect his diſplacement in 1640.—*Winthrop*, i. 299, 300; ii. 342-3.

In *Maſs. Rec.*, i. 308, is a liſt of perſons "propounded for magiſtrates" by the freemen, in October, 1640; with the number of votes by which they were reſpectively nominated.

[85] Before the act of 11 George IV.

and 1 William IV. c. 70, (1830,) Eaſter term of court began two weeks from the Wedneſday after Eaſter Sunday, and ended three weeks from the Monday following. The beginning of the term, varying as Eaſter fell earlier or later, ranged from April 8 to May 12; and the laſt Wedneſday, from April 29 to June 2.— By the Maſſachuſetts charter of 1691, the laſt Wedneſday of May was eſtabliſhed as the day of election.

oath, I doe not remember expreffed that ordinary faving, which is and ought to be in all oathes to other Lords, *Saving the faith and truth which I beare to our Soveraigne Lord the King,* though, I hope, it may be implyed.

Courts and Laws.
There are two generall Courts, one every halfe yeare, wherein they make Lawes or Ordinances: The Minifters advife in making of Laws, efpecially Ecclefiafticall, and are prefent in Courts, and advife in [86] fome fpeciall caufes criminall, and in framing of Fundamentall Lawes: But not many Fundamentall Lawes are yet eftablifhed: which, when they doe, they muft, by the words of their Charter, make according to the Laws of *England,* or not contrary thereunto.[87] Here they make taxes and levies.

[86] For "advife in," the M.H.S. MS. has, "in hearing."

[87] This was one of the "great reafons . . . which caufed moft of the magiftrates and fome of the elders not to be very forward in this matter" of fundamental laws; "for that it would profeffedly tranfgrefs the limits of our charter, which provide, we fhall make no laws repugnant to the laws of England, and *that we were affured we muft do.* But *to raife up laws* by *practice* and *cuftom* had been no tranfgreffion; as in our church difcipline," &c. (*Winthrop,* i. 323.) The "Body of Liberties," adopted December, 1641, was fo framed as, if poffible, to avoid this difficulty. They were "expreffed only under the name and title of *Liberties,* and not in the exact form of *Laws* or Statutes," and the General Court did not *enact* them, but "with one confent fully authorize and earneftly entreat all that are and fhall be in authority to *confider them as* laws," and not to fail to inflict punifhment for every violation of them. — *Lib.* 96; 3 *Mafs. Hift. Coll.,* viii. 236.

At a later period, when the afcendency of the parliament was eftablifhed, and Maffachufetts was for the time relieved from apprehenfion of the lofs of her charter, the General Court denied, with lefs referve, the authority of the laws of England. They "did ever honor the parliament, and were ready to perform all due

There are befides foure quarter Courts for the whole Jurifdiction, befides other petie Courts, one every quarter, at *Bofton*, *Salem*, and *Ipfwich*, with their feverall jurifdictions, befides every towne, almoſt, hath a petie Court for fmall debts, and trefpaffes under twenty fhillings.

In the generall Court, or great quarter Courts, before the Civill Magiftrates, are tryed | all actions and caufes civill and criminall, and alfo Ecclefiafticall, efpecially touching non-members: [88]And they themfelves fay, that in the generall and quarter Courts, they have the power of Parliament, Kings Bench, Common Pleas, Chancery, High Commiffion, and Star-chamber, and all other Courts of England, and in divers cafes have exercifed that power upon the Kings Subjects there, as is not difficult to prove. They have put to death, banifhed, fined men, cut off mens eares, whipt, imprifoned men, and all thefe for Ecclefiafticall and Civill offences, and without fufficient record. In the leffer quarter Courts are tryed, in fome, actions under ten pound, in *Bofton*,[8)] under twenty, and

Actions and caufes.

26

obedience, etc., to them according to our charter, etc.;" but they rebuked Dr. Childs and his fellow-petitioners in 1646, who "did impudently and falfely affirm, that we are obliged to thofe laws [of England] by our generall charter and oath of allegiance," (*Winthrop*, ii. 285, 288,) and they explicitly declared, "our allegiance binds us not to the laws of England any longer than while we live in England, for the laws of the parliament of England reach no further," &c.— *Ibid.* 288.

[88] The nine lines following (ending with "fufficient record") are not in the M.H.S. MS.

[8)] *Mafs. Rec.*, i. 169, 276.

all criminall caufes not touching life or member. [90] From the petie quarter Courts, or other Court, the parties may appeale to the great quarter Courts, from thence to the generall Court, from which there is no appeale, they fay: Notwithftanding, I prefume their Patent doth referve and provide for Appeales, in fome cafes, to the Kings Majefty.

Grand Juries. The generall and great quarter Courts are kept in the Church meeting-houfe at *Boflon*. Twice a yeare, in the faid great quarter Courts held before the generall Courts, are two grand Juries fworne for the Jurifdiction, one for one Court, and the other for the other, and they are charged to enquire and prefent offences reduced, by the Governour, who gives the charge, moft an-end, under the Heads of the *ten Commandments:* | [91] And a draught of

[90] Five lines following ("From the petie . . . Kings Majefty.") are not in the M.H.S. MS.—When Dr. Child and his fellow-petitioners demanded an appeal to England, in 1646, Gov. Winthrop "told them he would admit no appeal, nor was it allowed by our charter;" and the Court fuftained his judgment.—*Winthrop*, ii. 285, 290.

[91] Thirteen lines following (to the end of the paragraph) are not in the M.H.S. MS.

At leaft two draughts of a body of fundamental laws had been prefented to the General Court: one by Mr. Cotton, in October, 1636,—"a copy of Mofes his judicials, compiled in an exact method," (*Winthrop*, i. 202,)—which is perhaps the fame that was again prefented to the November Court in 1639, and printed in England in 1641, as "An Abftract of the Laws of New England as they are now eftablifhed," (repr., 1 *Mafs. Hift. Coll.*, v. 171-192.) and in a more complete form, by William Afpinwall, in 1655; and another, framed by Nathaniel Ward, prefented November, 1639, and, with Mr. Cotton's, referred by the Court to the governor and others "to confider of, and fo prepare it for the" May Court, 1640.—*Winthrop*, i. 322. Thefe "two models were digefted with divers alterations and additions, and *abbreviated*, and fent to every town

a body of fundamentall laws, according to the judiciall
Laws of the Jews, hath been contrived by the *Miniſters*
and *Magiſtrates*, and offered to the generall Court to be
eſtabliſhed and publiſhed to the people to be confidered
of, and this fince his Majeſties command came to them to
fend over their Patent:[92] Among which Lawes, that was
one I excepted againſt, as you may fee in the paper following, entituled, *Of the Church her liberties, prefented to
the Governour and Magiſtrates of the Bay,* 4. *Martii,*
1639.[93] Notwithſtanding, a by-law, to that or the like

(12), to be confidered of firſt by the
magiſtrates and elders, and then to
be publiſhed by the conſtables to all
the people," &c.—*Ibid.* Comp. *Maſs.
Rec.,* i. 379. Lechford was employed
to tranfcribe the "breviats of propofitions" to be fent to the towns, (see
p. 31, *poſt,*) and his Journal ſhows
that in January, 1639–40, he made for
the governor "a coppie of the Ab-
ſtract of the Lawes of New England,"
and numerous copies of "the Lawes
for the Country" and "the Breviat of
the body of Lawes," in January and
February. Mr. Ward wrote to Winthrop, Dec. 22, 1639: "Yf Mr. Lach-
ford have writt them out, I would be
glad to perufe one of his copies if I
may receive them."—4 *Maſs. Hiſt.
Coll.,* vii. 27. Mr. Gray, in his excellent paper on the Early Laws of Maffachufetts, in 3 *Maſs. Hiſt. Coll.,* viii.
attributing the compofition of the
Body of Liberties to Ward, (on the

authority of Winthrop, ii. 55,) remarks,
that it "exhibits throughout the hand
of the practifed lawyer, familiar with
the principles and the fecurities of
Engliſh Liberty." (p. 199.) Without
detracting from whatever honor may
be due to Mr. Ward for his firſt draught,
it is very poffible that while Lechford
was tranfcribing the much revifed and
amended "breviats," the "hand of a
practifed lawyer" left fome of its
traces on his work.

[92] The lords commiffioners for foreign plantations ordered, April 4,
1638, that the patent ſhould be fent
over to them by the firſt ſhip. The
demand was renewed the next year;
and a letter from Mr. Cradock, en-
clofing the order, was received by
Winthrop before the meeting of the
May Court.—*Winthrop,* i. 269, 274,
299; *Hubbard,* 268–271. See p. 34,
poſt, and note 103.

[93] See p. 31, *poſt.*

Tryals.

effect, hath been made,[94] and was held of force there when I came thence: yet I confesse I have heard one of their wiseft speak of an intention to repeale the same Law.

Matters of debt, trespasse, and upon the case, and equity, yea and of heresie also, are tryed by a Jury. Which although it may seeme to be indifferent, and the Magistrates may judge what is Law, and what is equall, and some of the chief Minifters informe what is heresie, yet the Jury may find a generall verdict, if they please; and seldome is there any speciall verdict found by them, with deliberate arguments made thereupon, which breeds many inconveniences.

The parties are warned to challenge any Juryman before he be sworn; but because there is but one Jury in a Court for tryall of causes, and all parties not present at their swearing, the liberty of challenge is much hindred,[95]

---

[94] March, 1635-6.—*Mass. Rec.*, i. 168.

[95] The hinderance was not entirely removed by the Body of Liberties, which secured the right of challenge in all cases; but the challenge was not to be allowed unless the other jurors, or the bench (as the challenger might elect), should find it "juft and reafonable."—3 *Mass. Hift. Coll.*, viii. 221.

There was another hinderance to the freedom of challenge which Lechford omits to mention. The wife of Francis Welton, of Salem, was brought before the church, in 1637, on the charge, that, "when a matter of difference between her and another was at the Court put unto the Jury, she excepted againft two of the jury men," (members of the fame church with herself,) "who were therefore offended, and with them others alfo;" and she was inftructed, that, although the law grants such exception in case of confanguinity or some nigh relation, the ground or reafon muft be showed to the judge of the court. And her allegation that one of the challenged

and fome inconveniences doe happen thereby. Jurors are returned | by the Marfhall, he was at firft called the Bedle[96] of the Societie. Seldome is there any matter of record, faving the verdict many times at randome taken and entred, which is alfo called the judgment. [97]And for want of proceeding duly upon record, the government is cleerely arbitrary, according to the difcretions of the Judges and Magiftrates for the time being. And humbly I appeale to his royall Majefty, and his honourable and great Counfell, whether or no the proceedings in such matters as come to be heard before Ecclefiafticall Judges, be not fit to be upon Record; and whether Regifters, Advocates, and Procurators, be not neceffary to affift the poore and unlearned in their caufes, and that according to the warrant and intendment of holy Writ, and of right reafon. I have knowne by experience, and heard divers have fuffered wrong by default of fuch in *New-England*. I feare it is not a little degree of pride and dangerous im-

28

---

jurors was "all one with the party againft her" appears to have been regarded as an aggravation of her offence. — *Rev. John Fifk's Notes,* in *Coll. Effex Inftitute,* i. 40, 41.

[96] The title of this officer was changed by the General Court, September, 1634, when James Penn, who had been appointed in 1630 "a beadle to attend upon the governor," &c., was granted a falary of £20 ; "his office to be Marfhall of the Court." — *Mafs. Rec.,* I. 74, 128. His fucceffor was Edward Michelfon, of Cambridge, appointed November, 1637. — *Ib.* 217.

[97] What follows, to the end of the paragraph, (twenty lines,) is not found in the M.H.S. MS., at the correfponding page, but is written in fhorthand, as a marginal note, on p. 9, of the MS., correfponding to pp. 12, 13, of the firft edition.

providence to flight all former lawes of the Church or
State, cafes of experience and precedents, to go hammer
out new, according to feverall exigencies; upon pretence
that the Word of God is fufficient to rule us: It is true,
it is fufficient, if well underftood. But take heede my
brethren, defpife not learning, nor the worthy Lawyers of
either gown, left you repent too late.[98]

[98] Lechford — forbidden "to plead any man's caufe except his own," and only efcaping the Court's cenfure by "promifing to attend his calling, and *not to meddle with controverfies,"* — had reafon to fpeak feelingly of the fmall efteem had for the legal profeffion in Maffachufetts. There was very little encouragement for the future in the "Body of Liberties," which permitted "every man that findeth himfelf unfit to plead his own caufe in any court," to employ any man againft whom the court doth not except, to help him, *provided he give him no fee or reward for his pains."* — *Lib.* 26; 3 *Mafs. Hift. Coll.,* viii. 220.

Mr. Cotton, in a fermon delivered early in 1640. took occafion, by way of "ufe." to drop a word of "reproof to unconfcionable Advocates;" such as "bolfter out a bad cafe by quirks of wit, and tricks and quillets of Law." . . . "And for men that profefs Religion (as many Lawyers do) to ufe their tongues as weapons of unrighteoufneffe unto wickednefs . . . to plead in corrupt Caufes, and to ftrain the Law to that purpofe, were I to fpeake in place where [fuch are?] I fhould think it meet to fpeak more." But, he remarks, "I have not I thinke fo much caufe to fpeak of it here, but in moft places of the world I might fpeak of it." — *Expos. of* 13*th Chap. of Revelation,* p. 163.

Mr. Ward, preaching before the General Court, in May, 1641, advifed "that magiftrates fhould not give private advice, and take knowledge of any man's caufe before it came to public hearing. This was debated after in the General Court;" but fome of the magiftrates oppofed the making a law to this effect, for divers reafons. "1. Becaufe we muft then provide lawyers to direct men in their caufes. . . . 4. [The private hearing by a magiftrate] prevents many difficulties and tedioufnefs to the court to underftand the caufe aright (*no advocate being allowed,* and the parties being not able, for the moft part, to open the caufe fully and clearly, efpecially in public)." — *Winthrop,* ii. 36.

The parties in all caufes, fpeake themfelves for the moſt part, and fome of the Magiftrates where they thinke caufe requireth, doe the part of Advocates without fee or reward. Moſt matters | are prefently heard, and ended the fame Court, the party defendant having foure dayes warning before; but fome caufes come to be heard again, and new fuits grow upon the old.

Profane fwearing, drunkenneffe, and beggers, are but rare in the compaffe of this Patent,[99] through the circum- fpection of the Magiftrates, and the providence of God hitherto, the poore there living by their labours, and great wages, proportionably, better then the rich, by their ſtocks, which without exceeding great care, quickly waſte.

Prophaneneſſe beaten downe.

A Paper of certaine Propoſitions to the generall Court, made upon requeſt, 8. Iunii, 1639.

1. IT were good, that all actions betweene parties, were entred in the Court book, by the Secretary, before the Court fits.

[99] "One may live there from year to year, and not fee a drunkard, hear an oath, or meet a beggar." — *New England's Firſt Fruits.* "In feven years, among thoufands there dwelling, I never faw any drunk, nor heard an oath, nor [faw] any begging, nor Sab- bath broken." — Hugh Peters, *Cafe impartially communicated,* &c. (1660). "I thank God, I have lived in a Colony of many thoufand Engliſh al- moſt thefe twelve yeares; am held a very fociable man; yet I may confi- derately fay, I never heard but one Oath fworne, nor never faw one man drunk, nor ever heard of three women Adultereſſes, in all this time, that I can call to minde." — Ward's *Simple Cobbler of Agawam* (1647) p. 67. "There are none that beg, in the Countrey." — Joffelyn, *Voyages,* 182.

2. That every action be declared in writing, and the defendants anfwer, generall or fpeciall, as the cafe fhall require, be put in writing, by a publique Notarie, before the caufe be heard.

3. The Secretary to take the verdicts, and make forth the judiciall Commands or Writs.

4. The publique Notarie to record all the proceedings in a fair book, and to enter executions of commands done, & fatisfactions acknowledged.

5. The fees, in all thefe, to be no more then in an inferiour Court of Record in *England*, and to be allowed by the generall Court, or Court of Affiftants.

The benefit hereof to the publique good.

1. IT will give an eafie and quick difpatch to all Caufes: For thereby the Court and Jury will quickly fee the point in hand, and accordingly give their verdict and judgment.

2. The Court fhall the better know, conftantly, how to judge the fame things; and it is not poffible, that the Judges fhould, alwayes, from time to time, remember clearly, or know to proceed certainly, without a faithfull Record.

3. The parties may hereby more furely, and clearly obtaine their right; for through ignorance and paffion, men may quickly wrong one another, in their bare words, without a Record.

4. Hereby fhall the Law of God and Juftice be duly adminiftred to the people, according to more certaine and unchangeable rules, fo that they might know what is the Law, and what right they may look for at the mouthes of all their Judges.

5. Hereby the Subjects have a great part of their evidences and affurances for their proprieties, both of lands and goods.[100]

[100] What appears to be the original draught of thefe propofitions, with numerous erafures and amendments, is in Lechford's Journal, p. 57. He had there added, under the head of "The benefit hereof" &c., a *sixth* confideration: —

"The people may alfo ufe the publique notary in divers cafes, to the eafe of the magiftrates, and for making feverall writings, etc."

Immediately after thefe propofitions, the writer made this note, in short-hand: —

"The Court was willing to beftow employment upon me, but they faid to me that they could not doe it for feare of offending the Churches, becaufe of my opinions. Whereupon I thought good to propofe unto them as followeth, over the leaf," — where he has inferted "Certaine Propofitions to the Generall Court, 11. 4. 1639," which are printed in the Introduction to this volume.

Though the Court did not fee fit to conftitute the office of public notary, and to give Lechford the employment which he fought, the value of his fuggeftions was appreciated, and at the next Seffion (September, 1639), orders were made for recording judgments, "with all the evidence," — "wills, adminiftrations and inventories, as alfo of the dayes of every marriage, birth and death," — "all men's houfes and lands," — and "all the purchafes of the natives." Lechford's fchifmatical opinions on prophefying in the churches and the poffibility of a coming Antichrift, continued to difqualify him for the public fervice; and the Court infured the orthodoxy of the records by appointing Steven Winthrop (the governor's fourth fon) "to record things," and the next year (October, 1640) chofe Emanuel Downing to enter all bargains and fales of land, &c., at Salem, and Samuel Symonds at Ipfwich. — *Mafs. Records*, i. 275-6, 306.

Plaine dealing,

31   A Paper touching the Church her liberties, delivered at
      Boston, 4. *Martii*, 1639.

To the Right Worshipfull the Governour, Deputy Governour, Councellers, and Assistants, for this Jurisdiction.

WHereas you have been pleased to cause me to transcribe certain Breviats of Propositions,[101] delivered to the last generall Court, for the establishing a body of Lawes, as is intended, for the glory of God, and the welfare of this People and Country; and published the same, to the intent that any man may acquaint you, or the Deputies for the next Court, with what he conceives fit to be altered or added, in or unto the said lawes; I conceive it my duty to give you timely notice of some things of great moment, about the same Lawes, in discharge of my conscience, which I shall, as *Amicus curiæ*, pray you to present with all faithfulnesse, as is proposed, to the next

---

[101] See before, p. 27, and note 91. Lechford's account-book and journal show that he delivered twelve copies of "the Lawes for the Country" in December, 1639; "Five copies more . . . by the direction of our Governor, 11. 8. 1639; seven of them (and the former) had 3 lawes more added:" "A coppie of the Abstract of the Lawes of New England d^d to the Governor, 11. 15. 1639:" [Was this Mr. Cotton's, printed under the same title in 1641?] "A coppy of the breviat of the body of Lawes for the Country, 12. 5. 1639:" "Three coppyes of the said breviat delivered to the Governor, besides the first, 12. 12. 1639:" "One coppy dd to Mr. B[ellingham?]:" "One coppy . . . delivered to Mr. Bellingham, wth one copy of the originall Institution and Limitation of the Counsell, at 4^s and 2^s, 12. 17. 1639:" and, near the end of the same month (February, 1640), "Seven coppyes more of the said breviate."

generall Court, by it, and the reverend Elders, to be further confidered of, as followeth:[102]

1. It is propounded to be one chiefe part of the charge, or office of the Councell intended, to take care that the *converſion* of the *Natives* be endeavoured.

2. It is propoſed, as a liberty, that a convenient *number of Orthodoxe Chriſtians, | allowed to plant together in this Juriſdiction, may gather themſelves into a Church, and elect and ordaine their Officers, men fit for their Church, yet I have heard Maſter *Cotton* ſay, that a Church could not be without the number of ſixe or ſeaven at leaſt, and ſo was their practiſe while I was there, at *Weymouth*, and *New Taunton*, and at *Lin*, for *Long Iſland*; becauſe if there are but three, one that is offended with another, cannot upon cauſe tel the Church, but one man.[1 2']

32

* Although ſome have held that three or two may make a

[102] The clauſe to which Lechford objects, and which agrees in ſubſtance with a proviſion of the order of March, 1636, (*Maſs. Rec.*, i. 168,) is not found in the Body of Liberties, as printed by Mr. Gray, 3 *Maſs. Hiſt. Coll.*, viii., from the MS. copy in the Athenæum. It may have been omitted on the reviſion of theſe laws in December, 1641, (*Winthrop*, ii. 55.) or on the ſubſequent reviſion made by order of the Court in 1644, by Winthrop, Dudley, and Hibbens. (*Maſs. Rec.*, ii. 61.) The "Declaration of the Liberties given to the Churches," (*Body of Lib.*, 96, §1,) aſſures to "all the people of God within this Juriſdiction who are not in a church way, *and be orthodox in judgment* . . . full libertie to gather themſelves into a Church eſtate. *Provided* they do it *in a Chriſtian way*, with due obſervance of the rules of Chriſt,"

etc. *Lib.* 58 declares that "the Civill Authoritie hath power and libertie to ſee the peace, ordinances and Rules of Chriſt obſerved in every Church, according to his word, ſo it be done in a civil and not in an eccleſiaſticall way."

In Welde's *Anſwer* to *W. Rathband*, (Lond., 1644,) he ſays, "There was a time in *New-England* (for ſome few years ſpace) before ſuch a law was made, and then Churches did uſe to gather without any notice given to Magiſtrates or other Churches. But after the opinions grew on, and experience diſcovered the danger, there was a law made that none muſt conſtitute any Church but firſt give notice thereof to Magiſtrates and Churches, and ſince that this courſe is duly obſerved." P. 32.

[102'] See Cotton's *Way of the Churches*, p. 53.

places, giving notice to feven of the next Churches, one moneth before thereof, and of their names, and that they may exercife all the ordinances of God according to his Word, and fo they proceede according to the rule of God, and fhall not be hindered by any Civill power; nor will this Court allow of any Church otherwife gathered.

> This claufe (*nor will the Court allow of any Church otherwife gathered*) doth as I conceive contradict the firft propofition.

My reafons are thefe.

1. If the *converfion of the Natives* muft be endeavoured, then fome wife and godly men (they fhould be of your *graveft and beft men*) muft bee fent forth to teach them to know God.

2. When fuch are fent, they muft bee either fent immediately by the Lord, or mediately by his Churches.

3. If the Churches fend men, they that are fent muft be fent by impofition of *hands of the Prefbyterie*.

> *Now when Churches are thus gathered or planted, they are gathered by Minifters, doing the works of Apoftles and Euangelifts, which hath ever been, and is the ordinary and regular way of gathering or planting Churches, (and not as is propofed, a convenient number of Orthodoxe Chriftians, gathering themfelves into a Church) and yet when fuch a Church is gathered by*

Church-meffengers and Minifters, this Court is advifed not to allow the fame; which, I conceive, is to fay, The converfion of the Natives fhall not be endeavoured, orderly, according to the rule of God.

Againe, it would be confidered, that when men are fent forth, whether they fhould not be fent forth *two*, and by *two* at leaft, as the Scriptures beare, and for divers good reafons, which lye not hid to your wifdomes.

That you would be pleafed to fhew unto the Elders thefe things to be confidered, and that they would well weigh, whether or no thofe Minifters and Meffengers fent by Churches, fhould not vifit the Churches which they plant?

Other things there are, wherein, I think, I could alfo, to good purpofe, move fomewhat to your Worfhips, which lyes more directly in the way, and calling, I have been educated, if I were required, but this thing lying upon my confcience, *I could not well paffe by: Wherefore I fhall requeft it may be confidered.*

1. Whether it be not fit to leave out, at leaft, | that contradictory claufe, *viz. Nor will this Court allow of any Church otherwife gathered?*

2. Whether it be not better to let the liberty run thus, in generall,

The holy Church of God fhall enjoy all her juft liberties?

34

A Paper intended for the honoured John Winthrop, *Esquire, late Governour. Boſton, Maii* 2. 1640.

IF you ſee a neceſſity of *baptizing* them without, If an ingagement of *Propagation* of the truth to the Infidel *Natives:*

Then conſider, whether by the Kings leave, ſome Churches may not be appointed to ſend their chiefe Paſtors, and other Miniſters, to doe ſuch works.

Alſo, with ſome kind of ſubjection, or acknowledgement of authority of the Miniſterie in *England*, if it be but by way of advice, which is cleare to me you may doe:

I make no doubt but in all things requiſite for the ſtate of the Country, they will yeeld you all faire liberties.

Nay, I am perſwaded, the Kings Majeſty will not ſend any unexperienced Governour to afflict, but make you Patentees againe, or at leaſt, after the manner of other Plantations, reſtoring not onely favour, but other benefits, whereof, under God, to us *Engliſhmen*, he is the Fountaine. The Kings Attorney did offer ſome of you this in my hearing, I meane, the renewall of your Patent.[103]

[103] When and where could this offer have been made, in Lechford's hearing? The demand for the return of the patent to England, made by the Commiſſioners for the plantations, April 4, 1638, was communicated to Gov. Winthrop by letter from Thomas Mewtis, clerk of the Council. The repetition of the demand, in 1639, with "threats of further courſe to be taken" in the event of non-compliance, was received in a letter from Mr. Cradock, and "not being delivered by a *certain meſſenger, as the for-*

Nay further, if you would invent, and devife what the
King may doe for the Country, you might obtaine.
The very converfion of thefe poore naked people,
which is very hopefull, and much prepared for *per acci-
dens*, or Gods owne providence, bringing good out of
evill, will rejoyce the hearts of all Chriftians in our deare
native Countrey, and here: and of it felfe (if there were
no other defirable things here, as bleffed be God there
are many) would caufe a continued confluence of more
people then you can tell well where to beftow for the
prefent.

The Fifhing trade would be promoted with authority.

Hereby would you give the greater teftimony to the
caufe of Reformation.

Hereby will you, under God and the King, make

---

mer was," no reply was returned, and precautions were taken that the commiffioners "could not have any proof that it was delivered to the governor." *Winthrop*, i. 269, 274, 299; *Hubbard*, 268-271; 4 *Mafs. Hift. Coll.*, vi. 129. Winthrop and the records are filent as to any fubfequent renewal of the demand, or any interview with "the King's Attorney." It is poffible that Lechford alludes to fomething which paffed in England, on the trial of the *quo war-ranto*, or after judgment was rendered againft the patentees, in 1636, and before his coming to this country.

Who was the "certain meffenger" to New England in 1638, we are not told. There is, at leaft, ground for conjecture, that John Joffelyn, Gent, who failed from London three weeks after the date of the Commiffioners' order, landed at Noddle's Ifland, and was the gueft of Maverick July 10, went to Bofton the day following, and "prefented his refpects to Mr. Winthrop the Governour, and to Mr. Cotton," and next morning failed for Black Point, in Maine, (*Voyages*, 1, 12, 20,) and who was in the intereft of Sir Ferd. Gorges, — was charged with this fervice.

Church-work, and Common-wealth work indeede, and examples to all Countryes.

You will enrich your Countries both, in ſhort time. The Heathens in time, I am perſwaded, will become zealous Chriſtians, then will they labour, get cloathes, and ſubſtance about them. In vaine doe ſome think of civilizing them, either by the ſword, or otherwiſe, till (withall) the Word of God hath ſpoken to their hearts: wherein I conceive great advice is to be taken.

For which purpoſe a Preſſe is neceſſary,[104] and may be obtained, I hope, ſo that wiſe men watch over it.

36   Conſider how poorely your Schooles goe on, | you muſt depend upon *England* for help of learned men and Schollers, bookes, commodities infinite almoſt.

No doubt but the King, this way, will make your authority reach even to the Dutch Southward, and to the French Northward. *New-England* indeede without fraction. A facile way, taking out the core of malice.

The converſion and ſubduing of a Nation, and ſo great a tract of ground, is a work too weighty for ſubjects any much longer to labour under without Royall aſſiſtance, as I apprehend, I think, in religious reaſon.

---

[104] The preſs was already obtained, thanks to the liberality and foreſight of Rev. Joſſe Glover.— *Winthrop*, i. 289. "Wee have a printery here," wrote Hugh Peters, Dec. 10, 1638, "and thinke to goe to worke with ſome ſpeciall things."— 4 *Maſs. Hiſt. Coll.*, vi. 99. The "Book of Pſalmes" bears the imprint of 1640, and muſt have been in preſs, if not completed, when this "paper" of Lechford's was written.

If any fhall fuggeft, that your Churches may fend forth men of their own authority;

Confider, if it may be done warrantably by the Word of God, as peradventure it may be fo.

Yet you will be in danger, *rebus fic ftantibus*, of great imputations.

        That you infringe *Regall power*,
        And Ecclefiafticall.

Wheron adverfaries will fure enough make fearfull worke.

And befides, fome reformations (under favour) have been too deep, at leaft for others to follow.

They were alfo unexperienced of miffion to convert Infidels.

Is *Geneva* without her faults? or *Holland*, *Rotterdam*, *Amfterdam*, without theirs? what experience have they of miffion, or ever had?

Now I befeech you grave Sir, doe you thinke it good, honourable, fafe, for us poore men here, | or for the Religion and Profeffors thereof in generall, in the whole world, that fuch as have the name of the moft zealous, fhould be the firft example of almoft utmoft provocation to our owne Soveraigne?

For my part, I difclaime *Parker*,[105]

---

[105] Robert Parker, whom Mather calls "in fome fort the father of all the non-conformifts in our age," and "one of the greateft fcholars in the

And encline to *Hooker, Iewel*,[106] as to government.

Great men have great burthens, therefore they have their counfels croffe, and fometimes they ufe them both.

You heare enough on the other fide,

Heare now this, on this, and the Lord guide your fpirit.

Odere Reges dicta, quæ dici jubent.[107]

Englifh nation," and whom Cotton (*Way of the Churches Cleared*, pt. i. p. 13) names firft of the "gracious faints and faithfull witneffes" through whofe teaching the Congregationalifts of New England received "the chief doctrines and practices of [their] way, fo far as it differeth from other reformed churches," was, in 1571, rector of North Binfleet, in Effex; next year, at Weft Henningfield, and fubfequently at Dedham, in the fame county. Sufpended from his miniftry for refufing to fubfcribe Whitgift's three Articles (1584), he removed to Wilton, in Wiltfhire. He was compelled to leave England in confequence of his publication, in 1607, of *A Scholafticall Difcourfe againft Symbolizing with Antichrift in Ceremonies*, etc. He went to Holland, — was for a time at Leyden, in the companionfhip of Dr. Ames and Henry Jacob, — afterwards at Amfterdam; and died about 1614. Two years after his death was publifhed his treatife *De Politeia Ecclefiaftica Chrifti et Hierarchia oppofita,* *libri tres.* (Francof. 1616, 4to.) In this he maintained the doctrine "that the vifible Church inftituted by Chrift and his Apoftles, to which the Keys are given, is not a *Diocefan*, or *Provincial*, or *National* Affembly, but a particular *Congregation*." (See Cotton's *Way Cleared*, pt. ii. p. 23.)

[106] Richard Hooker, author of the famous treatife *Of the Laws of Ecclefiaftical Polity*; and John Jewel, Bifhop of Salisbury, who wrote the not lefs famous *Apologia Ecclefiæ Anglicanæ*, firft publifhed in 1562. In his preface, (chap. i. § 2,) Hooker declares, "as [his] final refolute perfuafion: Surely the prefent form of church-government which the laws of this land have eftablifhed is fuch, as no law of God nor reafon of man hath hitherto been alleged of force fufficient to prove they do ill, who to the uttermoft of their power withftand the alteration thereof."—*Works*, (Keble, 4th ed.) i. 127.

[107] Seneca, *Œdipus*, act. 3. 520.

Newes from New-England.

Thefe are the Minifters of the Bay.

AT *Bofton*, mafter *Cotton* Teacher, mafter *Wilfon* Paftor: At *Roxbury*, mafter *Weld*[108] Paftor, mafter *Eliott* Teacher: At *Dorchefter*, mafter *Mather* Paftor or Teacher, and Mafter *Burgh*[109] out of office: At *Braintree*, Mafter *Thomfon* Paftor, Mafter *Flint* Teacher:[110] At *Weymouth*, Mafter *Newman* Paftor, Mafter *Parker* out of office:[111] At *Hingham*, mafter *Hubbard* Paftor, mafter

^(marginal note: Minifters names)

[108] Thomas Weld failed for England, on a miffion for the Colony, Aug. 3, 1641, in the fame veffel with Lechford, and did not return again to Maffachufetts. *Winthrop*, ii. 25, 31. Hubbard (371) says that "he had given the greateft encouragement of any man elfe for invitation of his friends to come over to New England, yet it was obferved true of him, which fome note of Peter the hermit, who founded an alarum and march to all other Chriftians, to the Holy Land, but a retreat to himfelf."
In October, 1645, the General Court "defired [his] prefence here, and fpeedy return." (*Mafs. Rec.*, ii. 137.) The defire was not gratified. Mr. Weld became the minifter of St. Mary, Gatefhead, Co. Durham, and died in England, 23d March, 1661. — Savage's *Geneal. Dict.*

[109] Jonathan Burr came to New England in 1639, and, after having been received a member of the church in Dorchefter, was called to office as the affiftant of Richard Mather. Before the call was accepted, fome unguarded expreffions, "favoring of familifm," awakened apprehenfions of Mr. Burr's orthodoxy, and "the *Evil One*, difturbed at the happinefs of Dorchefter, very ftrongly endeavoured a *Mifunderftanding* between Mr. Mather and Mr. Burr, and the *Mifunderftanding* did proceed fo far as to produce a *Paroxifm*." — *Magnalia*, b. iii. pt. 2. c. 3. Of the treatment and happy compofure of the paroxyfm, fee the *Magnalia*, l. c., and *Winthrop*, ii. 23. Mr. Burr died Aug. 9, 1641, (a few days after Lechford failed from Bofton.) He was a graduate of Corpus Chrifti College. Cambridge, and, before coming to New England, had been a minifter at Rickingfhall [Rickinghall,] co. Suffolk. — Savage's *Geneal. Dict.*

[110] See before, p. 15, note 42.

[111] See before, p. 22, and note 78. The Rev. Samuel Newman fucceeded Thomas Jenner at Weymouth, and

Peck Teacher: They refuse to baptize old *Ottis* grandchildren, an ancient member of their own Church.[112] At *Charlestowne*, Master *Symms* Pastor, master *Allen* Teacher: At *Cambridge*, master *Sheppard* Pastor, master *Dunster*

became pastor of the church, which, after one or more unsuccessful trials, was at last "gathered with approbation of magistrates and elders," January 30, 1639.— Mr. James Parker, "a godly man and a scholar," had lived in Dorchester, removed to Weymouth, and was a deputy from that town to the General Court, 1639-42. He afterwards preached for some years at Portsmouth, but was not settled in the ministry there.— See *Winthrop*, ii. 93; Savage's *Geneal. Dict.*

[112] Of Rev. Peter Hobart, "a man well qualified with ministerial abilities, though not so fully perfuaded of the Congregational discipline as some others were," (as Hubbard judged, p. 192,) see the memoir in the *Magnalia*, b. iii. pt. 4. c. 1., and Savage's note to *Winthrop*, ii. 223.

Robert Peck was a graduate of Magdalen College, Oxford, (A. M., 1603,) and had been minister at Hingham, co. Norfolk, for more than thirty years before he came to New England in 1638. He was ordained teacher of the church in our Hingham, Nov. 8, 1638; but returned home in 1641, resumed his rectory in old Hingham in 1646, and died there

in 1656.— *Winthrop*, i. 275; *Magnalia*, b. iii. pt. 4. c. 1; Blomefield's *Hist. of Norfolkshire*, ii. 424, 425; Brook's *Lives*, iii. 263.

John Otis, born about 1581, at Glastenbury, co. Somerset, came to New England in 1635, and was one of the settlers of Hingham in that year. His daughter Hannah, wife of Thomas Gill, had *two* children baptized the same day, January, 1644, one of whom may have been born before Lechford wrote. Another daughter, Margaret, wife of Thomas Burton, had a daughter baptized May 30, 1641.— Savage's *Geneal. Dict.; N. E. Geneal. Reg.*, v. 223. A few years afterwards this Thomas Burton, "a sojourner, and of no visible estate in the country," was a figner, with Dr. Child, Fowle, Maverick, and others, of that famous petition to the General Court for the redress of sundry grievances, one of which was that of being, "themselves and their children, debarred from the seals of the covenant, except they would submit to such a way of entrance and church covenant, as their consciences would not admit." See *Winthrop*, ii. 261, 262; Hutchinson's *Collection*, 188-196.

Newes from New-England.

School-mafter;[113] divers young Schollers are there under him to the number of almoft twenty: At *Watertowne*, mafter *Phillips* | Paftor, mafter *Knolls* Paftor:[114] At *Dedham*, another mafter *Phillips*[115] out of office, and mafter *Allen* Paftor or Teacher: At *Sudbury*, mafter *Brown*[116] in office, mafter *Fordham*[117] out of office: At *Lynne*, maf-

[113] See after, pp. 52, 53. Before *Plaine dealing* was publifhed, "Mafter *Dunfter*, School-mafter," became Prefident of Harvard College.

[114] Rev. John Knowles, "a godly man and a prime fcholar," was ordained paftor, at Watertown, Dec. 9, 1640, as colleague of Mr. Phillips: "and fo they had now two paftors and no teacher, differing from the practice of the other churches," &c. — *Winthrop*. ii. 18.

[115] Rev. John Phillip, who had been rector of Wrentham (co. Suffolk), in England, and had married a fifter of the great puritan divine, Dr. William Ames, was at Dedham in 1638. He declined feveral invitations to the work of the miniftry in Maffachufetts, and returned home in the autumn of 1641, — failing from Bofton, Oct. 27, in the fhip with John Humfrey, Rev. Robert Peck, and two other minifters. He was a member of the Weftminfter affembly of divines. — *Winthrop*, ii. 86; Savage, *Geneal. Dict.*; Calamy (*Contin.*), ii. 797.

Rev. John Allin, the firft minifter of Dedham (ordained April 24, 1639), had alfo been a preacher at Wrentham in England. — Savage, *Geneal. Dict.* His "virtues and merits," fays Mather, "were far from the fmalleft fize, among thofe who 'did worthily in Ifrael.'" — *Magnalia*, b. iii. pt. 2. ch. 22.

[116] Rev. Edmund Browne came to New England in 1637 or '38, and was ordained minifter of the church gathered at Sudbury, in Auguft, 1640. By a power of attorney which he executed in July, 1639, it appears that he was then of Watertown, and had married the widow of John Loverum (or Loveran). — Lechford's *MS. Journal*, 87.

[117] Robert Fordham, who was for a fhort time at Cambridge, and afterwards at Sudbury, removed to Long Ifland, where, early in 1644, he was a principal planter at Hempftead, and is firft named in the patent for that townfhip granted by Gov. Kieft, in November of the fame year. He was fubfequently fettled at Southampton. — Savage, *Geneal. Dict.*; Brodhead's *Hift. of N. York*, i. 388; *Doc. Hift. of N. Y.* (8vo.) iii. 189.

ter *Whiting* Paſtor, maſter *Cobbett* Teacher: At *Salem*, maſter *Peter* Paſtor,[118] maſter *Norris* Teacher, and his Sonne a Schoole-maſter: At *Ipſwich*, maſter *Rogers* Paſtor, maſter *Norton* Teacher, and maſter *Nathaniel Ward*, and his ſonne,[119] and one Maſter *Knight*, out of employment: At *Rowley*, Maſter *Ezek. Rogers* Paſtor, Maſter *Miller:* [120]

[118] Hugh Peters ſailed for England in the ſame ſhip with Lechford, Aug. 3, 1641. Rev. Edward Norris, ordained teacher at Salem, March 18, 1640, remained in office there until his death, Dec. 23, 1659. — Savage, *Geneal. Dict.* His ſon Edward was ſchool-maſter from 1640 to 1676, and died in 1684. Rev. John Fiſk, who had taught a ſchool at Salem, and occaſionally preached there between 1637 and 1640, removed to Wenham before Lechford wrote. — *Ibid.*

[119] John Ward, educated at Emmanuel College, Cambridge, came to New England in 1639, aſſiſted Mr. Rogers, at Rowley, for a ſhort time, and in the winter of 1639-40, with his brother-in-law, Giles Firmin, was projecting the ſettlement of a plantation at Pentuckett or Cochichawick. March 23, 1641, Thomas Gorges wrote from Acomenticus (York, Me.) to Gov. Winthrop: "We have ſent younge Mr. Ward of Newbury a call. I hope the Lord will be aſſiſting to us in it;" and Winthrop tells how, in the ſpring of 1641, Ward, going from Paſca-taquack to Acomenticus, with Hugh Peter and Mr. Dalton, loſt his way, and "wandered two days and one night without food or fire." He returned to Maſſachuſetts, ſettled at Haverhill before 1642, and, on the gathering of the church there, was ordained paſtor, in October, 1645. — *Winthrop,* ii. 29, 252; *Mass. Rec.* i. 290: Hutch. *Collection,* 108; 4 *Mass. Hist. Coll.,* vii. 274, 334; Chaſe, *Hist. of Haverhill,* 37, 39, 58. (See after, p. 45.)

"Maſter *Knight*" is not named in the Hiſt. Society's MS. This was Rev. William Knight, who had lately come to Ipſwich, where he had a grant of land in 1639. He began to preach at the New Meadows (Topsfield), in 1641, but returned to England before 1648. — Savage, *Geneal. Dict.*

[120] John Miller, a graduate of Cambridge (A.B. 1827), was at Dorcheſter in 1636; afterwards, of Roxbury; miniſter at Rowley, 1639, as aſſiſtant or colleague of Mr. Rogers, and firſt town clerk there. He was deſignated by the elders in 1642, (with George

Newes from New-England.

At *Newberry*, Mafter *Noyfe* Paftor, Mafter *Parker* Teacher:[121] He is fonne of Mafter *Robert Parker*, fometime of *Wilton*, in the County of *Wiltes*, deceafed, who in his life time writ that mif-learned and miftaken Book *De Politeia Ecclefiaftica*.[122] At *Salifbury*, Mafter *Worfter*[123] Paftor: At *Hampton*, Mafter *Bachellor*[124] Paftor, Mafter

Phillips and William Tompfon,) for the miffion to Virginia, but did not accept the call. He fubfequently removed to Yarmouth, where he preached for a fhort time, but appears to have been living at Roxbury again in 1647, and died at Groton, June 12, 1663. — *Winthrop*, ii. 78 ; *Hubbard*, 410 ; Johnfon, *W. W. Providence*, b. ii. c. 11 ; Savage, *Geneal. Dict.*

[121] Thomas Parker, and his coufin James Noyes, had taught in the fame fchool in Newbury (co. Berks), England ; "came over in *one Ship;* were paftor and teacher of *one Church;* and Mr. Parker continuing always in Celibacy, they lived in one *Houfe*, till death feparated them for a time." — Rev. Nich. Noyes, in *Magnalia*, b. iii. pt. 2. c. 25.

[122] See before, p. 37, note 105.

[123] William Worfter was the firft minifter of the new plantation "begun upon the north fide of Merrimack, called Sarisbury, now (1639) Colchefter," — ordered "henceforward to bee called Salsbury," by the General Court in October, 1640. *Winthrop*, i. 289 ; *Mafs. Rec.*, i. 305. He is fuppofed to have come to New England in 1639 ; was admitted freeman May 13, 1640 ; and died Oct. 28, 1663. — Savage's *Geneal. Dict.*

[124] Winthrop records the arrival of "old Mr. Batchelor, being aged 71," a fellow-paffenger with Thomas Welde, June 5, 1632 ; and elfewhere gives account of his troubles at Lynn ; his unfuccefsful attempt to eftablifh a plantation at Mattakeefe (Yarmouth), in 1637 ; his fall and penitence at Hampton, in 1641 ; and of the ftop put by the General Court, to the gathering of a new church at Exeter, in 1644, to which he was to be called as paftor (i. 78, 176, 260 ; ii. 44, 177, 211). From 1647 to 1650, he was at Portfmouth ; returned to England in 1653 or '54 ; and died at Hackney, near London, in 1660, in the one hundredth year of his age. See Savage's note to *Winthrop*, i. 78 ; *Mafs. Rec.*, i. 100, 103, 236 ; Lewis, *Hift. of Lynn* (2d ed.) 78, 92-97. Several of his letters are printed from the Winthrop papers, in 4 *Mafs. Hift. Coll.*, vii. 88-109.

Mr. Bachiler, and his colleague at Hampton, — Rev. Timothy Dalton, —

86   *Plaine dealing*,

Dalton Teacher: There are other School-maſters which I know not, in ſome of theſe townes.[125]

The Magiſtrates in the Bay *are theſe:*

Magiſtrates names.

Maſter *Bellingham* the preſent Governour, maſter *Endecot* the preſent Deputy Governour, maſter *Winthrop*, maſter *Dudley*, maſter *Humfrey*, maſter *Saltonſtall*, maſter *Bradſtreat*, maſter *Stoughton*, maſter *Winthrop junior*, maſter *Nowell*, Aſſiſtants. Maſter *Nowell* is alſo Secretarie. Maſter *Stephen Winthrop* is Recorder, whoſe office is to record all Judgments, Mariages, Births, Deaths, Wills and Teſtaments, Bargaines and Sales, Gifts, Grants, and Mortgages.[126] There is a *Marſhall*,[127] | who is as a *Sheriffe* or *Bailiffe*, and his Deputy is the *Gaoler*[128] and executioner.

39

Marriages. Teſtaments. Adminiſtrat. Burials.

Marriages are ſolemnized and done by the Magiſtrates,[129]

were by no means ſo well agreed as were the two miniſters of Newbury. See *Winthrop*, ii. 45, 177, and 4 *Maſs. Hiſt. Coll.*, vii. 102.

[125] Lechford omitted to name the miniſters of *Concord*, where the eminent Peter Bulkley was teacher, and John Jones paſtor. Mr. Jones removed to Fairfield, Conn., with ſeveral of his pariſhioners, in September, 1644. — Savage, *Geneal. Dict.;* *Winthrop*, i. 167, 189, 217; ii. 73.

[126] See before, p. 30, note 100.

[127] See before, p. 28, note 96.

[128] Richard Bracket was "appointed to keepe the priſon," and "to bee at the commandment of the magiſtrates for any ſpecial ſervice," by the General Court, November, 1637. In 1639, his ſalary was increaſed to £20 per annum. — *Maſs. Rec.*, I. 217, 260.

[129] John Robinſon (in *A Juſtification of Separation*, &c., p. 352) refers to "almoſt twenty ſeverall ſcriptures [cited in his *Apology*, ch. vi.], and nine diſtinct reaſons grounded upon them, to prove that the celebra-

and not by the Minifters. *Probats of Teftaments, and granting of Letters of Adminiftration, are made and granted in the generall or great quarter Courts. At Burials, nothing is read, nor any Funeral Sermon made, but all the neighbourhood, or a good company of them,

*Caufes touching Matrimonie, and Teftaments, and other Ecclefiafticall caufes, have been anciently by the good lawes of England, committed to the Clergie, upon

better grounds then many are aware of. Brethren, I pray confider well that the Apoftle doth allow judgements of controverfies to the Church, 1 Cor. 6. And fo they did anciently in other countries, as well as in England, as appeares by S. Auguftines profeffion thereof, cited by one lately, viz. That he (the faid Father, and other holy men of the Church) *fuffered the tumultuous perplexities of other mens caufes touching fecular affaires, either by determining them by judging, or in cutting them off by entreaties: Which labour* (faith he) *we endure with confolation in the Lord, for the hope of eternall life. To which moleftations, the Apoftle tyed us, not by his owne judgement, but by his judgement who fpake in him.* Befides, fhould they judge thefe things, and labour for, and watch over us in the Lord, and not be recompenced as long as they doe well? I speak not to countenance undue exactions, bribes, or other corruptions. I intend brevity, and therefore make bold to refer my Reader to the many learned arguments both in Law-books and Divinity of this fubject.

tion of marriage, the buryall of the dead, *are not ecclefiafticall actions, apperteyning to the miniftry, but civill, and fo to be performed;*" and argues that "the proper works of the miniftry muft needes be workes of religion," while "thefe are civill duties, and fo practifed by the fervants of God in all ages." In Maffachufetts, it was not deemed advifable "to make a *law*, that marriage fhould not be folemnized by minifters," becaufe this would be "repugnant to the laws of England;" but due care was taken "to bring it to a *cuftom*, by practice, for the magiftrates to perform it." See *Winthrop*, i. 323, and comp. ii. 313. So, in Plymouth, the firft marriage (May 12, 1621), "according to the laudable cuftome of ye Low-Countries, . . . was thought moft requifite to be performed by the magiftrate, as being a civill thing," &c., and Brad-

ford (101) adds, "this practifs hath . . . been followed by all the famous churches of Chrift in thefe parts to this time, — Anno 1646." Mr. Savage has "difcovered no record of a marriage performed by a clergyman prior to 1686, except in Gorges' Province, by a clergyman of the Church of England." — *Proceed. Mafs. Hift. Soc.*, 1858-60, p. 283.

The publication of the intention, or "contraction," of marriage, was, however, fometimes folemnized by a fermon. Mather alludes to this as "the old ufage of New England," when fpeaking of Mr. Cotton's fermon, in 1651, at the contraction of Rev. Samuel Danforth to the daughter of Mr. Wilfon. — *Magnalia*, b. iv. c. 3. § 6. A MS. note-book of Henry Wolcot, of Windfor, preferves the heads of a fermon by Rev. John Warham, Nov. 17, 1640, "at the

come together by tolling of the bell, and carry the dead folemnly to his grave, and there ftand by him while he is buried. The Minifters are moft commonly prefent.[130]

contracting of Benedict Alvord, and Abraham Randall." (The former married Joan Newton, Nov. 26; the latter, Mary Ware, Dec. 1.) That this difcourfe was *practical* and pointed may be inferred from the felection of the text, — *Ephes.* vi. 10, 11, ("Finally, my brethren, be ftrong in the Lord," &c. "Put ye on the whole armour of God, that ye may be able to ftand againft the wiles of the devil.") — and from one of the "ufes" to which it was applied by the preacher, — "to teach us, *that the ftate of marriage is a warfaring condition.*"

Robert Baylie, the Presbyterian divine, in *A Diffuafive from the Errors of the Time* (Lond., 1646), alleging that, "whatever crotchets the *Brownifts* have fallen into, the *Independents* punctually do follow the moft and worft of them," says, "Firft, for the marriage blefling, . . . they fend it from the Church to the Town-houfe; making its folemnization the duty of the magiftrate: this is the conftant practice of all in New England. The prime of the Independent Minifters now at London, have been married by the Magiftrate: and all that can now be obtained of them, is, to be content that a Minifter, in the name of the Magiftrate and as his commiffioner, may folemnize the holy band." — pp. 115, 116.

[130] "Concerning burials, this they fay: all prayers either over or for the dead, are not only fuperftitious and vain, but alfo are idolatry, and againft the plain fcriptures of God . . . Mourning in black garments for the dead, if it be not hypocritical, yet it is fuperftitious and heathenifh: funeral fermons, they alfo utterly condemn, becaufe they are put in the place of trentals, and many other fuperftitious abufes follow thereby. To be brief . . . the Nonconformifts will have the dead to be buried in this fort, (holding no other way lawful,) namely, that it be conveyed to the place of burial, with fome honeft company of the Church, without either finging or reading, yea, without all kind of ceremony heretofore ufed, other than that the dead be committed to the grave, with fuch gravity and fobriety as thofe that be prefent may feem to fear the judgments of God, and to hate fin, which is the caufe of death; and thus do the beft and right reformed churches bury their dead, without any ceremonies of praying or preaching at them." — J. Canne's *Neceffitie of Separation* (1634); *Hans. Knolly's Soc. ed.*, pp. 112, 113. Comp. Mather's *Ratio Difciplinæ*, 117.

They are very diligent in traynings of their fouldiers *Trainings of Mufters.* and military exercifes, and all except Magiftrates, and Minifters beare armes, or pay for to bee excufed, or for fpeciall reafons are exempted by order of Court. The Captains and officers are fuch as are admitted of the Church.

But the people begin to complain, they are ruled like *Grievances.* flaves, and in fhort time fhall have | their children for the moft part remain unbaptized: and fo have little more priviledge then Heathens, unleffe the difcipline be amended and moderated.

It is feared, that Elections cannot be fafe there long, *Danger.* either in Church or Common-wealth. So that fome melancholy men thinke it a great deale fafer to be in the midft of troubles in a fettled Common-wealth, or in hope eafily to be fettled, then in mutinies there, fo far off from fuccours.

At *New Plymouth* they have but one * Minifter, mafter *Rayner;*[131] yet mafter *Chauncey* lives there, and one mafter *Smith*, both Minifters, they are not in any office

New Plymouth Patent.
M. *Rayner.*
M. *Smith.*
* Eccles. 12, 11. One fhepheard:

James 3. 1. *Not many mafters.* Whether this be their ground, I know not; but what ever there be in others to advife and affift, the deciding, determining voice, I meane alfo the negative, in fome cafes, ought, as I think, to be in the Paftor: Be there never fo many Minifters in the Church, *Doe nothing without your Paftor or Bifhop*, faith *Irenæus*; for whatfoever is faulty in the Church, the Bifhop is firft and principally blamed, *Rev.* 2. and 3.

The firft inftance, as far as is known, of prayer at a funeral, in Maffachufetts, was at the burial of the Rev. William Adams, of Roxbury, Auguft 19, 1685, when, as Judge Sewall noted in his Diary, " Mr. Wilfon, minifter of Medfield, prayed with the company before they went to the grave."— Palfrey's *Hift of N. E.*, iii. 495, note 1.

131 John Reyner, born at Gilderfome, Yorkfhire, was educated at

> *Plaine dealing,*

<small>M. *Chauncey* his controverſie.</small> there;[132] maſter *Chauncey* ſtands for dipping in baptiſme onely neceſſary, and ſome other things, concerning which there hath been much diſpute, and maſter *Chauncey* put to the worſt by the opinion of the Churches adviſed withall.[133]

<small>Taunton. M. *Hooke*. M. *Streate*, their ordination.</small> *Cohannet*, alias *Taunton*, is in *Plymouth* Patent. There is a Church gathered of late, and ſome ten or twenty of the Church, the reſt excluded. Maſter *Hooke* Paſtor, maſter *Streate* Teacher.[134] Maſter *Hooke* received ordi-

Magdalen College, Cambridge. He came to New England about 1635; the next year ſucceeded Rev. Ralph Smith at Plymouth, and continued in the miniſtry there until 1654. — *Bradford*, 351; Davis's *Morton's Memorial*, 216, 217; Savage's *Geneal. Dict.*

[132] Ralph Smith came with Higginſon in 1629, and after brief ſtay at Nantaſket, removed to Plymouth, where he was "kindly entertained and houſed, . . . and exerciſed his gifts amongſt them, and afterwards was choſen into yᵉ miniſtrie, and ſo remained for ſundrie years." *Bradford*, 263. In the winter of 1635–36, when John Norton was preaching in Plymouth, and there was hope of effecting his ſettlement there, Mr. Smith "layed downe his place of miniſtrie, partly by his owne willingnes, . . . partly at the deſire and by yᵉ perſuaſion of others." — *Ibid*, 351; comp. *Winthrop*, i. 175.

Charles Chauncy had been at Plymouth ſince 1638. He left there, in the latter part of 1641, to become paſtor at Scituate, where he remained until choſen preſident of Harvard College in 1654. — See *Bradford*, 382–84; *Winthrop*, i. 330; *Magnalia*, b. iii. pt. 2, c. 23; I *Mafs. Hiſt. Coll.*, x. 171; and the ample memoir in Fowler's *Memorials of the Chaunceys*, 1–37.

[133] See *Bradford*, 383; *Winthrop*, i. 330–31.

[134] William Hooke (A.M. Trin. Col., Oxford, 1623) is named as a landholder in Taunton in May, 1639. Inquiry for the date of his ordination there is hopeleſs, ſince Mr. Savage has "aſked in vain." *(Geneal. Dict.)*. He removed to New Haven, and was ordained teacher there about 1644; thence he returned to England, in 1656, to become the domeſtic chaplain of Cromwell. — Bacon's *Hiſt. Diſcourſes*, 62–73; Savage, in note to *Winthrop*, i. 251; Emery's *Miniſtry*

nation from the hands of one mafter *Bifhop* a School-mafter, and one *Parker* an Husbandman,[135] and then mafter *Hooke* joyned in ordaining mafter *Streate*. One mafter *Doughty*, a Minifter, oppofed the gathering of the Church there, alleadging that according to the Covenant of *Abraham*, all mens children that were of baptized parents, and fo *Abrahams* children, ought to be baptized; and fpake fo in publique, or to that effect, which was held a difturbance, and the Minifters fpake to the Magiftrate to order him: The Magiftrate commanded the Conftable, who dragged mafter *Doughty* out of the Affembly. He was forced to goe away from thence, with his wife and children.[136]

M. *Doughty* his controverfie.

41

of *Taunton*, i. 63-155; Baylies' *Hift. of Plymouth Colony*, i. 290-95.

Nicholas Street removed to New Haven, where he was elected and ordained teacher (Nov. 26, 1659) as Mr. Hooke's fucceffor, and the affociate of Davenport. After the latter removed to Bofton, in 1668, Mr. Street remained fole minifter of the New-Haven Church till his death, April 22, 1674. — Bacon's *Hift. Difcourfes*, 155-57; Baylies' *Plymouth*, i. 295.

[135] "Mafter Bifhop" was, probably, John Bifhop, afterwards minifter of Stamford, Conn. See *N. E. Geneal. Reg.*, viii. 156. Trumbull (*Hift. of Conn.*, i. 286) fays, the meffengers of the Stamford Church, fent to feek a minifter, "travelled on foot, through the wildernefs, to the eaftward of Bofton, where they found Mr. John Bifhop, who left England before he had finifhed his academical ftudies, and had completed his education in this country."—William and John Parker, probably brothers, were among the purchafers of Taunton, in 1637. The latter was a reprefentative in the General Court, in 1642 and 1643. Baylies' *Plymouth*, ii. 2, 282; Savage, *Geneal. Dict.* John Parker and John *Bufhop* (as the name is recorded,) of Taunton, were propounded for freemen, June 1, 1641. *Plym. Col. Rec.*, ii. 17.

[136] In the earlier draught was added: "And being a man of eftate when he came [to] the country, is undone."

Divers other Towns and Ministers.

There are also in this Patent divers other Plantations, as *Sandwich, Situate, Duxbury, Greenesharbour*,[137] and *Yarmouth*. Ministers there are, master *Leveridge*,[138] mas-

M. H. S. MS. In July and August, 1639, Lechford was attorney for Elizabeth, sister of Francis Doughty and wife of William Cole, in a suit to recover from her brother a share of their deceased father's estate, and her promised marriage portion: and it was "for going to the jewry and pleading w*th* them out of Court," in this cause, or another between the same parties, that Lechford was debarred by the Court "from pleading any man's cause hereafter," &c., as his MS. journal shows. See *Mass. Rec.*, i. 270; ii. 205, 206.

Francis Doughty was the son of a merchant of the same name, of Bristol, who died before 1637. In a recognizance for appearance at the next Quarter Court, made in July, 1639, he is styled, of Dorchester. In March, 1641, then of Taunton, he was fined by the Plymouth Court for selling powder to the Indians. (*Plym. Col. Rec.*, ii. 8.) In August, 1639, his sister, in a petition to the General Court of Massachusetts, averred that he "had a purpose to remove his dwelling forth out of the jurisdiction of this Court; where, this complainant cannot tell." (Lechford's *Journal*.) He was, afterwards, at Rhode Island, where he made brief stay; and, in 1641, betook himself to the Dutch at Manhattan, from whom he and his associates procured, March 28, 1642, a patent for Mespath (since, Newtown, L. I.). He failed, however, "to secure the happy home" which (Mr. Brodhead tells us) he came, from persecution in Massachusetts, to seek : for he was fined and imprisoned by Kieft, — "threatened with this and that" by Stuyvesant, — obliged to quit Mespath for Flushing, — and driven from Flushing to Virginia. See Brodhead's *Hist. of New York*, i. 333, 367, 411, 472.

[137] Incorporated as a township, March, 1641, and named *Rexhame*, but, within a year afterwards, called by its present name, *Marshfield*. *Plym. Col. Rec.*, xi. (Laws) 37. The Rev. Richard Blinman, with the friends who came with him to New England, settled first at Green's Harbour, probably in 1640 : but he had left that place (and Plymouth Colony) before Lechford's book was written. See after, p. 54.

[138] The Rev. William Leveridge, or Leverich (A. M. Emman. College, Cambridge, 1629), after successive removals from Dover to Boston, and from Boston to Duxbury (where he was for a short time the assistant of Rev. Ralph Partridge), settled at Sandwich before 1640, and was teach-

ter *Blackwood*,[139] mafter *Mathews*,[140] and mafter *Andrew Hallet*,[141] a School-mafter. Mafter *Saxton* alfo, who was comming away when we did.[142]

At the Ifland called *Aquedney*,[143] are about two hun- [Ifland *Aquedney*.] dred families. There was a Church, where one mafter

er of the church there. For what further is known of him and his work, fee Savage's *Geneal. Dict.*, and note in *Winthrop*, i. 115; Freeman's *Hift. of Cape Cod*, ii. 38.

[139] Chriftopher Blackwood was for a fhort time at Scituate, after the removal of the Rev. John Lothrop to Barnftable in 1639. He returned to England in 1642. — Deane's *Scituate*, 172, 222.

After the name of Mr. Blackwood, Lechford had inferted (in the M.H.S. MS.) that of "Mr. Thomas." This was probably William Thomas, of Marfhfield, who is fuppofed to have come to New England with Mr. Blinman. On a fubfequent page (54) will be found mention of "a broyle betweene one Mafter Thomas . . . and Mafter Blindman," which refulted in the removal of the latter from the colony.

[140] Marmaduke Matthews preached at Yarmouth from 1639 to 1643. Of him and his many troubles, fee Frothingham's *Charleftown*, 121-29; Freeman's *Cape Cod*, ii. 180, 182; Johnfon's *W.W. Providence*, b. iii. c. 7.

[141] Andrew Hallet removed from Lynn to Sandwich in 1637, and to Yarmouth about 1640. — Savage's *Geneal. Dict.*

[142] The M.H.S. MS. adds, "And I know not what ftayed him, he is very aged and white." The Rev. Peter Saxton (A.M. Trin. Col., Cambridge, 1603), whom Mather calls "a ftudious and a learned perfon, a great Hebrician," was at Scituate in 1640, but did not long remain there. He was probably one of the four minifters who returned to England, Oct. 27, 1641, in the fhip with John Humfrey. — See *Magnalia*, b. iii. pt. 4, c. 1; *Winthrop*, ii. 85; Brook's *Lives*, iii. 139; Savage's *Geneal. Dict.*

Of the "worthy inftruments" whom Morton, s. a. 1542, names "among the fpecialeft" in Plymouth colony, Lechford omits the Rev. John Lathrop (Scituate, 1634-39; Barnftable, 1638-53); Rev. John Mayo, Mr. Lathrop's colleague at Barnftable (ord. April 15, 1640); and Rev. Ralph Partridge, firft minifter of Duxbury (1637-58).

[143] "The Ifland," — a name fpecially appropriated to Rhode Ifland by the Englifh who firft planted there. With the locative or objective affix, *Aqueduct*, or *Aquidnick*, fignifying,

Clark was Elder:[144] The place where the Church was, is called *Newport*, but that Church, I heare, is now diffolved;[145] as alfo divers Churches in the Country have been broken up and diffolved through diffention. At the other end of the Ifland there is another towne called *Portfmouth*, but no Church: there is a meeting of fome men, who there teach one another, and call it Prophefie. Thefe of the Ifland have a pretended civill government of their owne erection, without the Kings Patent.[146] There lately they whipt one mafter *Gorton*,[147] a grave man, for

Mafter Gorton whipt and banifhed.

to, on, or *at* the Ifland. [Thus, in Acts xxvii. 16, Eliot wrote, "*ahquednet hettamun Clauda*," for "an ifland called Clauda.] The *diminutive* of this name, Aqued*enc*fet, or Aquid*nef*ick, was given to "the little Ifland in the mouth of the Bay" (4 *Mafs. Hift. Coll.*, vi. 267).

The M.H.S. MS. reads here,— "about *one* hundred families : There *is* one church," etc.

[144] For "Elder," the M.H.S. MS. has "Paftor," and adds : "There is Mr. *Lenthall* a minifter out of office and imployment, and lives very poorly. Mr. *Doughty* alfo is come to this Ifland. The place where the church is, is called *New porte*." To the name of Mr. Lenthall is this note, in the margin : "He ftood upon his minifterie and againft the Church Covenant in the Bay, and diverfe joyneing to choofe him their minifter at Wey-

mouth, by fubfcribing to a paper for that end, he was cenfured in the genrall Court at Bofton, and fo were they that joyned in that election, and one of them named *Brittaine* for words faying that fome of the Minifters in the Bay were Brownifts, and that they would not [*fic*] till it came to the fwords point, was whipt, and had eleven ftripes." Comp. p. 22, *ante*, and fee notes 77–80.

[145] "But that church ... through diffention." Thefe three lines are not in the M.H.S. MS., where the fentence ends with "*Newport*."

[146] The words "pretended civil" are not in the M.H.S. MS.

[147] Of Samuel Gorton,—whofe hiftory Mr. Savage has, in a fingle line, reduced to its effence : "a moft active religious difturber of feveral places,"—fee the Memoir by Judge Staples, prefixed to "Simplicities De-

denying their power, and abufing fome of their | Magif- 42
trates with uncivill tearmes; the Governour, mafter *Coddington*, faying in Court, *You that are for the King, lay hold on Gorton;* and he againe, on the other fide, called forth, *All you that are for the King, lay hold on Coddington;*[148] whereupon *Gorton* was banifhed the Ifland: fo with his wife and children he went to *Providence*.[149] They began about a fmall trefpaffe of fwine, but it is thought fome other matter was ingredient.[150]

fence," *R. I. Hift. Coll.*, ii.; Mackie's *Life of Gorton*, in Sparks's *Amer. Biography; Notice* (by Chas. Deane) in *N. E. Geneal. Regifter*, iv. 201-21; Savage's *Geneal. Dict.*, and note to *Winthrop*, ii. 58; Arnold's *R. Ifland*, i. 163, *et feq.*

"The fum of the prefentment of Samuel Gorton, at Portfmouth" was printed by Edward Winflow in his reply to Gorton (*Hypocrifie Unmafked*, Lond., 1646), pp. 54, 55, — whence it is copied by Arnold, *Hift. of R. Ifland*, i. 170-72. The precife date of his trial or of his banifhment from the Ifland has not been afcertained. Mr. Arnold (i. 172) places it in March or April, 1640; Dr. Palfrey (*Hift. of N. E.*, ii. 119) writes "1640 or 1641;" Mr. Savage, in his *Geneal. Dictionary*, and Mr. C. Deane, in *N. E. Geneal. Regifter*, iv. 206, are filent.

148 Comp. the prefentment (14th count) in *Hypocrifie Unmafked*, 55; Arnold's *Hift. of R. I.*, i. 172.

149 Roger Williams wrote to Gov. Winthrop, from Providence, under the date, "8. 1ft 1640," fome account of Gorton's proceedings there: "Mafter Gorton having foully abufed high and low at Aquednick, is now bewitching and bemadding poor Providence, . . . denying all vifible and external ordinances, in depth of Familifm, againft which I have a little difputed and written, and fhall (the Mott High affifting.) to death. As Paul faid of Afia, I of Providence; (almoft) all fuck in his poifon, as at firft they did at Aquednick," etc. — *Hypocrifie Unmafked*, 55; *Arnold*, i. 172. Comp. *N. E. Geneal. Reg.*, iv. 216. Mr. Arnold reads the date of this letter, Oct. 1, 1640; Mr. Deane and Dr. Palfrey (ii. 120), March 8, 1641. That the latter is the true reading hardly admits of doubt.

150 Winflow exprefly ftates that Gorton's punifhment was "all for breach of the civil peace and notori-

Plaine dealing,

New Providence. At *Providence,* which is twenty miles from the said Island, lives master *Williams,* and his company of divers opinions; most are Anabaptists; they hold there is no true visible Church in the *Bay,* nor in the world, nor any true Ministerie.[151] This is within no Patent, as they say; but they have of late a kind of government also of their owne erection.[152]

ous contempt of authority, without the least mention of any points of religion, on the Government's part. *Hyp. Unm.* The "small trespass" was by "an ancient woman's cow going into the field where Samuel Gorton had some land."

[151] See *Winthrop,* i. 293, 367. After his re-baptism by Holliman, Roger Williams "walked in the Baptists' way about three or four months," then "brake from the society, and declared at large the ground and reason of it, — that their baptism could not be right, because it was not administered by an Apostle. After that, he set himself upon a way of *Seeking,*" etc. (Rich. Scott's letter, in App. to Fox's *N. E. Fire-Brand Quenched,* 247.) "He advised [the church at Providence] to forego all, to dislike everything, and wait for the coming of new Apostles; whereupon they dissolved themselves, and became that sort of sect which we term *Seekers,* . . . owning of no true *Churches* or *Ordinances* now in the world." *Magnalia,* b. vii. c. 2, § 7; comp. Callender's

Hist. Disc., 109, 110. When Mr. Williams was in England, in 1643 and '44, he made numerous converts to his peculiar belief, and the sect of *Seekers* — of whom Baxter called him the father — became considerable in numbers and influence. Robert Baillie, the Presbyterian divine, wrote from London, in June, 1644, that the Independents "are divided among themselves. One Mr. Williams has drawn a great number after him, to a singular independency, denying any true church in the world." Not long afterwards, he wrote again : " Sundry of the Independents are stepped out of the church and follow my good acquaintance, Mr. Roger Williams, who says there is no church, no sacraments, no pastors, no church-officers, or ordinance in the world, nor has been since a few years after the Apostles." Hanbury's *Memorials,* ii. 439, 444. Of the sect of Seekers, see *Reliq. Baxterianæ,* pt. i. 76 ; Edwards's *Gangræna,* pt. ii. p. 131.

[152] This alludes probably to the articles of agreement entered into by

One mafter *Blakefton*,[153] a Minifter, went from *Bofton*, having lived there nine or ten yeares, becaufe he would not joyne with the Church; he lives neere mafter *Williams*, but is far from his opinions.

There are five or fix townes, and Churches upon the River *Conneƈticot*,[154] where are worthy mafter *Hooker*, mafter *Warham*, mafter *Hewet*, and divers others, and mafter *Fenwike* with the Lady *Boteler*,[155] at the rivers

M. *Blakefton*.

Connecticot.

Lady *Boteler.*

the inhabitants of Providence, July 27, 1640. " It was but a flight departure from the primitive democracy, flill it forms an era in our colonial hiftory, and for feveral years conftituted the town government." — Arnold's *Hift. of R. Ifland*, i. 108; fee Staples, *Annals of Prov.*, 40–43.

153 In the M.H.S. MS., this name is written *Blackftone*. Of the little which is known of the Rev. William Blackflone, the firft known white fettler of Bofton, and the "earlieft permanent civilized refident" in what now conftitutes the State of Rhode Ifland, ample exhibition may be found in Blifs's *Hift. of Rehoboth*, 2-14; Daggett's *Hift. of Attleborough*, 29; Callender's *Hift. Difcourfe* (Elton's ed.), App. v.; S. Davis, in 2 *Mafs. Hift. Coll.*, x. 170; Savage's *Geneal. Diƈt.*, and note on *Winthrop*, i. 44, 45; Drake's *Bofton*, 95–97; Arnold's *R. Ifland*, i. 99, ii. 568–70. He lived on the traƈt afterwards called the "Attleborough Gore" (included in Rehoboth north purchafe, in 1661), near the river fince called by his name, in the fouth part of what is now the townfhip of Cumberland, R. I.

154 Lechford had not vifited Conneƈticut, and his notice of the river towns is brief and unfatisfaƈtory. He names Mr. Hooker of Hartford, but omits his colleague, the Rev. Samuel Stone. John Warham was paftor, and Ephraim Huet teacher, of the church at Windfor. It does not appear that Wethersfield had an organized church before 1641, when Henry Smith is fuppofed to have been ordained there. Peter Prudden (who was fettled over the church of Milford, April, 1640) and Richard Denton (who removed to Stamford early in 1641) had preached, perhaps only occafionally, at Wethersfield, before Mr. Smith's ordination. — Comp. Trumbull's *Hift. of Conn.*, i. 108, 120, 121, 279, 280; *Winthrop*, i. 305, 306; Goodwin's *Foote Genealogy*, Introd., xxxviii.

155 George Fenwick of Brinkburn, co. Northumberland, a gentleman of good family and eftate, was interefted

mouth in a faire houfe, and well fortified, and one mafter *Higgifon*, a young man, their Chaplain.[156] Thefe plantations have a Patent; the Lady was lately admitted of mafter *Hookers* Church, and thereupon her child was baptized.

43   The Lady *Moody*[157] lives at *Lynne*, but is of *Salem*

in the Earl of Warwick's grant to Lord Say and Sele and others, ufually termed the "old patent" of Connecticut. In July, 1639, as the reprefentative of the grantees, he came, "with his lady and family, to make a plantation at Saybrook," "landing at New Haven from the firft fhip that ever caft anchor in that place." (*Winthrop*, i. App. A. 59, 60, and p. 306; *Hutchinfon*, i. 97.) His wife, Alice, was a daughter of Sir Edward Apfley of Thackham, co. Suffex, Knt., and, at the time of her marriage with Mr. F., the widow of Sir John *Boteler* of Teflon, co. Kent. She died at Saybrook, in November or December, 1645, fhortly after the birth (Nov. 4) of her fecond daughter, Dorothy. The elder daughter, Elizabeth, whofe baptifm Lechford mentions, married, in England, after her father's deceafe, her coufin, Roger Fenwick, Efq., younger brother of Major John F., the proprietor of Salem Tenth, New Jerfey. See *Will of Geo. Fenwick*, in *Conn. Rec.*, i. App. vii.; Berry's *Suffex Genealogies* (Apfley); Burke's *Extinct Baronetcies* (Boteler, of Teflon.) Mr. Fenwick, having tranf-

ferred to Connecticut his intereft in the fort and plantation at Saybrook, returned to England foon after the death of his wife, and entered the fervice of the parliament. In 1648, he "was made a colonel and governor of Tynemouth caftle" (*Winthrop*, ii. App. A. 72), and was named a member of the high court of juftice for the trial of Charles I., but did not take his feat in that tribunal. In 1652, he was Governor of Berwick; and, the fame year, married Katharine, daughter of Sir Arthur Hafelrigge. He died at Berwick, March 15, 1657; and a monument in the church there honors his memory with the epitaph, "A good man is a public good."

[156] John Higginfon, eldeft fon of the Rev. Francis; afterwards affiftant and fucceffor of Rev. Henry Whitfield at Guilford, and, later, the honored minifter of the Salem church, 1660–1708. Before his fettlement at Guilford, — probably before his chaplaincy at Saybrook, — he taught fchool for fome time in Hartford, while profecuting his ftudies for the miniftry under Mr. Hooker.

[157] I have found no earlier mention

Newes from New-England.

Church, fhee is (good Lady) almoft undone by buying Lady *Moody*.
mafter *Humphries* farme, *Swampfcot*,[158] which coft her
nine, or eleven hundred pounds.

Beyond *Connecticott* are divers plantations, as *New-* *New Haven.*
Haven, alias *Quinapeag*, where mafter *Davenport* is Paftor, and one mafter *Iames*[159] a Schoole-mafter; and

of Lady Deborah Moody than in a memorandum by Lechford, acknowledging the receipt of £1. 11. "of my Lady Moody," April 22, 1639, — in payment of profeffional fervices probably. She had a grant of four hundred acres of land from the General Court, May 13, 1640. *Mafs. Rec.*, i. 290. Winthrop (ii. 123), writing in the fummer of 1643, tells how, "being taken with the error of denying baptifm to infants," fhe was admonifhed by the Salem Church, and removed to the Dutch colony, "to avoid further trouble, etc." By permiffion of Gov. Kieft, fhe fettled at Gravefend, L. I. See Brodhead's *Hift. of New York*, i. 367, 411.

[158] In the eaftern part of Lynn; granted to Mr. Humfrey by the General Court, May, 1635, and laid out, 1637-8, "betwixt the Clifte and the Foreft River near Marblehead." *Mafs. Rec.*, i. 147, 226. [The name "Swampfcot" — contracted from *Wannafquompfkut* — fignifies, at the cliff, upright rock, or rock-fummit.]

[159] Thomas James, ordained paftor at Charleftown, Nov. 2, 1632, was difmiffed in confequence of "fome occafions of difference" with his colleague, Mr. Symmes, as Winthrop relates, under date of March 11, 1636. He removed, after brief ftay at Providence, to New Haven, where he had a grant of land, Nov. 3, 1639, and was admitted a freeman June 11, 1640. See *Winthrop*, i. 94; 127, 182, ii. 95; Frothingham's *Charleftown*, 70-72; *N. Haven Col. Rec.*, i. 24, 35; Bacon's *Hift. Difc.*, 57-59. A greater than Mafter James was teaching at New Haven when Lechford wrote, or very fhortly afterwards. Ezekiel Cheever, "the father of New-England fchool-mafters," came with Davenport and Eaton in 1638. In February, 1642, a free fchool was eftablifhed in New Haven, and provifion made by the General Court for its fupport, "according to which order, 20£. a year was paid to Mr. Ezekiel Cheever, ... for two or three years," and his falary was increafed to £30. in Auguft, 1644. — *N. H. Col. Rec.*, i. 62, 210; Bacon's *Hift. Difc.*, 318-20.

another where master *Whitfield*[160] is: and another where master *Pridgeon*[161] is, and some others,[162] almost reaching to the Dutch plantation southward. Among these are my old acquaintance, master *Roger Ludlow*,[163] master

[160] Manunkatuck — named *Guilford*, July 6, 1643 — was purchased and planted, in 1639, by the Rev. Henry Whitfield and his associates. Mr. W. arrived in New Haven in July, 1639, in the ship with George Fenwick. He returned to England in 1650. — *N. H. Col. Rec.*, i. 96, 199; *N. E. Geneal. Reg.*, ix. 149; Trumbull's *Hist. of Conn.*, i. 207, 285, 295.

[161] For *Prudden*. Wepowaug, afterwards named *Milford*, was purchased of the Indians, Feb. 12, 1639, by Rev. Peter Prudden and his associates. Their church was gathered at New Haven, and Mr. P. was ordained their minister, April 18, 1640. He remained with them till his death, July, 1656, — "a man of great zeal, courage, wisdom, and exemplary gravity in his conversation." — Hubbard's *Hist. of N. E.*, 328. See also, *Magnalia*, b. iii. pt. 2, c. vi.; Bacon's *Hist. Discourses*, 55.

[162] These were *Rippowams* (named Stamford, April 6, 1642), settled in 1641, under the jurisdiction of New Haven; *Pequonnock*, or *Cupheag*, (Stratford), and *Uncowa* (Fairfield), begun to be settled in 1639, the former admitted to town privileges by Connecticut the same year; the latter in April, 1640. — Trumbull's *Conn.*, i.

121, 109, 110; *N. H. Col. Rec.*, i. 45, 58; *Conn. Rec.*, i. 35, 36, 41, 53.

[163] Mr. Ludlow probably accompanied Captains Mason and Stoughton in pursuit of the fugitive Pequots to Safco swamp, in June, 1637, and had thus an opportunity of exploring that fine tract of land which Capt. Stoughton pronounced to be "before Pequot, or the Bay either, abundantly." (Letter, in *Winthrop*, i. App. D.) In October, 1639, he had taken some steps toward the establishment of a plantation at Uncowa (Fairfield), and removed thither not long afterwards. *Conn. Rec.*, i. 35, 53. Considering the important position which Roger Ludlow held in two colonies, and the trusts with which he was honored, it is surprising that so little of his personal history and family relation has come to light. That little may nearly all be seen in Savage's *Geneal. Dict.*, and note to *Winthrop*, i. 28, and in Trumbull's *Connecticut*, i. 217, 218. A prefatory note to Mr. Brinley's admirable reprint of the Conn. Laws of 1673 gives reason for doubting the correctness of Dr. Trumbull's statement (l. c.), adopted by Mr. Savage, that Ludlow "removed with his family to Virginia." The fact of his return to England is placed beyond

Froſt,[164] fometime of *Nottingham*, and his fonnes, *Iohn Grey* and *Henry Grey;* the Lord in his goodneſſe provide for them; they have a Miniſter, whofe name I have forgotten, if it be not maſter *Blackwell*.[165] I do not know what Patent thefe have.

Long Iſland is begun to be planted, and fome two Miniſters are gone thither, or to goe, as one maſter *Peirſon*,[166]

[164] William Froſt was an early fettler at Uncowa, where he died in 1645. His will, of Jan. 6, 1644-5, is printed in *Conn. Rec.*, i. 465. John and Henry Grey were living in Boſton in 1639. Before May of that year, John married Elizabeth, daughter of William Froſt, and widow of John Watfon. Aug. 1, he fold a houfe and home-lot in Lynn, and, before September 28, removed (perhaps accompanying his father-in-law) to Uncowa, or the vicinity. Henry, the younger brother, had a houfe-lot granted in Boſton, Feb. 12, 1639. In an inſtrument executed Sept. 7, 1639, he is defcribed as "now of Boſton, heretofore citizen and merchant of London." He married Lydia, another daughter of William Froſt, after May, 1639. In September, 1640, he and his wife conveyed their houfe in Boſton to Thomas Lechford, in truſt, to be fold for their account. (Lechford's *MS. Journal.*) He foon afterwards followed his brother and wife's father to Uncowa, where he became a man of fome importance; was a deputy in 1656 and '57; and died in 1658.

[165] Adam Blakeman (as in his autograph now before me, but more often written by his contemporaries and defcendants, *Blackman*) became, in 1640, the firſt miniſter of Pequonnock (Stratford), where he continued to reſide until his death in 1665. — See the *Magnalia*, b. iii. pt. 2, c. 7; Trumbull's *Hiſt. of Conn.*, i. 280, 463; Savage's *Geneal. Dictionary.*

[166] Rev. Abraham Pierfon, from Yorkſhire (A.B. Trin Col., Cambr., 1632), came to New England in 1640, and was chofen miniſter of the church gathered at Lynn in November of that year, for removal to Long Iſland. — See *Winthrop*, ii. 6; Savage's *Gen. Dict.;* Trumbull's *Conn.*, i. 148.

Plaine dealing,

and mafter *Knowles*,[167] that was at *Dover*, alias *Northam*. A Church was gathered for that Ifland at *Lynne*, in the *Bay*, whence fome, by reafon of ftraitneffe, did remove to the faid Ifland; and one mafter *Simonds*, heretofore a fervant unto a good gentlewoman whom I know, was one of the firft Founders.[168] Mafter *Peter* of *Salem* was at the gathering, and told me the faid mafter *Henry Simonds* made a very cleare confeffion. Notwithftanding he yet dwels at *Bofton*, and they proceed on but flowly. The Patent is granted to the Lord *Starling;* but the *Dutch* claime part of the Ifland, or the whole: for their

---

[167] Sept. 28, 1641, James Farrett, agent of the Earl of Stirling, recorded at Bofton his formal proteft againft Edward Tomlyns and Timothy Tomlyns, "with *one Hanfard Knowles* and others, who have lately entered and taken poffeffion of fome parts of the Long Ifland," etc. See note to *Winthrop*, ii. 4. Thefe were of the company from Lynn and Ipfwich which went to Long Ifland in the fummer of 1641, and "finding a very commodious place for plantations, but challenged by the Dutch, they treated with the Dutch governor to take it from *them*," and obtained from Kieft a grant (June 6) of all the privileges they defired, on "very fair terms." The Maffachufetts Court "were offended at this, and fought to ftay them, not for going from us, but for ftrengthening the Dutch, our doubtful neighbors, and taking that from them which our king challenged and had granted ... to the Earl of Stirling." Some of the leaders, called before the October court, "were convinced and promifed to defift." *Winthrop*, ii. 34; Brodhead's *Hift. of New York*, i. 332-33. If Mr. Knollys actually went with this company to Long Ifland, he did not long remain there, for we know that he arrived in London, Dec. 24, 1641. Brook's *Lives*, iii. 492; *Winthrop*, ii. 28 (and note).

[168] Henry Symonds came to New England in July or Auguft, 1640, (failing from Briftol in the " Charles," or her confort, the " Hopewell "). He was admitted an inhabitant of Bofton Jan. 30, 1643, and died there in September of the fame year.—Lechford's *MS. Journal;* Snow's *Bofton*, 124, 125; Drake's *Bofton*, 278, *n*.

plantation is right over againſt, and not far from the South end of the ſaid Iſle. And one Lieutenant *Howe* pulling downe the Dutch Arms on the Iſle, there was like to be a great ſtir, what ever may become of it.[169] The Dutch alſo claime *Quinapcag*, and other parts.

At *Northam*,[170] alias *Paſcattaqua*, is maſter *Larkham* Paſtor. One maſter *H. K.*[171] was alſo lately Miniſter there, with maſter *Larkham*. They two fell out about baptizing children, receiving of members, buriall of the dead; and the contention was ſo ſharp, that maſter *K.* and his party roſe up, and excommunicated maſter *Larkham*, and ſome that held with him: And further, maſter *Larkham* flying to the Magiſtrates, maſter *K.* and a Captaine[172] raiſed Armes, and expected helpe from the

44

Paſcattaqua.

M. *Larkham* excommunicated.

A broyle or riot.

[169] See *Winthrop*, ii. 4-7; Brodhead's *Hiſt. of New York*, i. 297-99. The "great ſtir" was quieted by the interchange of letters in Latin, by Kieft and Dudley.

[170] Dover was for a ſhort time called Northam, after a pariſh of that name near Bideford, co. Devon, where the Rev. Thomas Larkham had been miniſter. Of the ſtrife between Mr. Larkham and Mr. Knollys, Winthrop gives a full account, ii. 27, 28. See alſo two letters from Hugh Peters, in 4 *Maſs. Hiſt. Coll.*, vi. 106, 107; Belknap's *New Hampſhire*, i. 46-49. Mr. Larkham failed for England in 1642, and Winthrop (ii. 92) gives a good reaſon for thinking "it was time for him to be gone." He became the miniſter of Taviſtock, Devonſhire, and notwithſtanding the evil report which followed him acroſs the Atlantic, he was honored as "a man of great ſincerity, ſtrict piety and good learning." Palmer's *Calamy*, i. 407. Edwards, in the *Gangræna* (1646: pt. 3. p. 97), gives brief and bitter notice of "one Maſter Larkin," who was then preaching ſomewhere in Kent, — "a fierce Independent."

[171] "Hanſard Knowles." — *M.H.S. MS.*

[172] "Captaine Underhill." — *Ibid.* Comp. *Winthrop*, ii. 27, where the

Bay; mafter *K.* going before the troop with a Bible upon a poles top, and he, or fome of his party giving forth, that their fide were *Scots,* and the other *Englifh :* [173] Whereupon the Gentlemen of Sir *Ferdinando Gorges* plantation came in, and kept Court with the Magiftrates of *Pafcattaqua,* (who have alfo a Patent) being weake of themfelves. And they fined all thofe that were in Armes, for a Riot, by Indictment, Jury, and Verdict, formally.[174] Nine of them were cenfured to be whipt, but that was fpared. Mafter *K.* and the *Captain* their leaders, were fined 100.l. a piece, which they are not able to pay. To this broyle came mafter *Peter* of *Salem,* and there gave his opinion, at *Northam,* that the faid excommunication was a nullity.[175]

Epifcopacie.

45   Mafter *Thomas Gorgs* fonne of *Captain Gorgs* of *Batcombe,* by *Chedder* in *Somerfetfhire,* is principall Commif-

Province of Maigne.

[173] "Knollys's calling his party *Scots,* and the other party *Englifh,* will be underftood when it is remembered that the battle of Newburn-upon-Tyne had been lately fought." — Palfrey's *Hift. of N. E.,* i. 591.

[174] "Mr. Larkham and his company ... fent to Mr. [Francis] Williams, who was governour of thofe in the lower part of the River [at Portfmouth and Dover], who came up with a company of armed men and befet Mr. Knolles' houfe where Capt. Underhill then was, ... and in the mean time they called a Court, and Mr. Williams fitting as judge, they found Capt. Underhill and his company guilty of a riot, and fet great fines upon them," etc. —*Winthrop,* ii. 28.

[175] See after, p. 53, where Lechford mentions this vifit of Hugh Peters to Dover as one of the "occurrences touching Epifcopacie."

captain is faid to have gathered his neighbors "to *defend himfelf,* and to fee the peace kept," Mr. Larkham having previoufly "laid violent hands upon Mr. Knolles."

sioner for the *Province of Maigne*, under Sir *Ferdinando*, but he was not at that Court at *Northam* himselfe.[176] Master *Wards* sonne[177] is desired to come into the *Province of Maigne*. There is one master *Ienner*[178] gone thither of late. There is want of good Ministers there; the place hath had an ill report by some, but of late some good acts of Justice[179] have been done there, and divers

[176] Thomas Gorges arrived at Boston in the summer of 1640, commissioned a member of the council for Maine, and its secretary. Winthrop found him "well disposed," and "careful to take advice of our magistrates how to manage his affairs." He remained a few days in Boston, and went to Maine in season to be present at the second meeting of the General Court for the province, September 8. In 1641, when Acomenticus was incorporated as a town, by charter of Sir Ferd. Gorges, his "well-beloved cousin" Thomas was named mayor; and he was also constituted deputy-governor of the province. *Winthrop*, ii. 9, and Savage's note; Sullivan's *Maine*, App. VI.; Williamson's *Maine*, i. 283-5; Hazard's *State Papers*, i. 47; *Letters of T. Gorges* to *Winthrop*, in 4 *Mass. Hist. Coll.*, vii. 333, 335.

[177] See before, p. 38, note 119.

[178] Rev. Thomas Jenner, who had been at Roxbury in 1634 or 1635; afterwards at Weymouth, where he preached for some years, and his name appears as deputy to the General Court, in May, 1640. In January, 1641, he was at Saco, commended thither by Winthrop, Humfrey, and other friends in Massachusetts; and Richard Vines (who was an Episcopalian) wrote that "he liked Mr. Jenner his life and conversation, and also his preaching, if he would let the Church of England alone." He was yet at Saco in April, 1646, though already "on the wing of removal;" returned to England, and was living in Norfolkshire in 1651. — *Winthrop*, i. 250 (and note), 287-88; Folsom's *Saco and Biddeford*, 81-83; *Mass. Hist. Coll.*, 3d Ser. iv. 144; 4th Ser. vii. 340, 341.

[179] The "good acts of justice" to which Lechford specially alludes, were, probably, the proceedings against the notorious George Burdett, late governor and preacher at Dover, and more recently at Acomenticus, where Thomas Gorges "found all out of order, for Mr. Burdett ruled all." In 1640, he was complained of and fined, on three several convictions, for gross misconduct, and soon afterwards re-

Plaine dealing,

Gentlemen [180] there are, and it is a Countrey very plentiful for fiſh, fowle, and veniſon.

Exeter. Not farre from *Northam* is a place called *Exeter*, where maſter *Wheelwright* hath a ſmall Church.[181]

Cape Anne. Fiſhing. And at *Cape Anne*, where fiſhing is ſet forward, and ſome ſtages builded,[182] there one maſter *Raſhley* is Chap-

turned to England. — See *Winthrop*, i. 276, 281, 291; ii. 10; *Hubbard*, 221, 353, 361.

[180] We may read here, with the M. H. S. MS., "divers *well accompliſht and diſcreete* Gentlemen there are."

[181] Rev. John Wheelwright, "being baniſhed from us, gathered a company, and ſat down by the falls of Paſcataquack, and called their town Exeter." *Winthrop*, i. 290. The ſettlement was commenced in 1638, and, Oct. 4, 1639, thirty-five planters ſubſcribed a combination for civil government, independent of other juriſdiction. *Hazard*, i. 463; Belknap's *N. Hampſhire*, i. ch. 1.

[182] "A fiſhing trade was begun at Cape Ann by one Mr. Maurice Tomſon, a merchant of London; and an order was made [by the General Court, in May, 1639], that all ſlocks employed in fiſhing ſhould be free from public charge for ſeven years." *Winthrop*, i. 307; *Maſs. Rec.*, i. 256, 257-8. Mr. Thompſon, if he came at all to New England, did not remain long. He was an enterpriſing mer-

chant, who was largely intereſted in trade with Canada, Virginia, the Weſt Indies, and Guinea; much employed by the company of Providence Iſland, the Virginia company, and the proprietors of the Somers Iſland, between 1632 and 1650; a member of the Guinea company; and, in 1653, one of the commiſſioners for governing the Somers Iſlands. See Sainſbury's *Calendar of Colon. Papers*, i. 151, 155, 294, 316–19, &c. Land was appropriated in his name at Cape Ann, and "Mr. Thomſon's frame" (probably for curing fiſh) is mentioned in the Glouceſter town records in 1650, as having formerly ſtood upon a "parcel of land in the harbour." Babſon's *Hiſt. of Glouceſter*, 50. Oſmond Douch and Thomas Milward (or Millard) were partners in the fiſhing buſineſs at Cape Ann in July, 1639, and the latter deſcribes himſelf as "of Cape Ann," in Auguſt, 1640. Lechford's *Ms. Journal*. They were probably employed by Mr. Thompſon, and were under the immediate direction of his agent, Samuel Maverick of Noddle's Iſland.

lain:[183] for it is farre off from any Church: *Rafhley* is admitted of *Bofton* Church, but the place lyeth next *Salem*, and not very far further from *Ipfwich*.
The *Ifle* of *Shoales* and *Richmonds Ifle*, which lie neere *Pafquattaqua*, and [184] good fifhing places.   Ifle of *Shoales* and *Richmond*.

About one hundred and fifty leagues from *Bofton* Eaftward is the *Ifle of Sables*, whither one *Iohn Webb*, alias *Evered*, an active man, with his company are gone with commiffion from the *Bay*, to get Sea-horfe teeth and oyle.[185]   Ifle of *Sables*.

Eaftward off *Cape Codd* lyeth an Ifland called *Martins*   *Martins* Vineyard.

---

[183] Thomas Rafhley was admitted to the Bofton Church, March 8, 1640, then called a "ftudent." He was at Exeter in 1646; returned to England, and was minifter at Bifhop Stokes, Hants; afterwards, it is faid, in Wiltfhire. *Geneal. Dict.* In 1641, Rev. Richard Blinman, with a part of the company who followed him from Wales, removed from Green's Harbor (Marfhfield) to Cape Ann, and gave to the plantation the name of Gloucefter.

[184] For "and" read "are." *M.H.S. MS.* Richmond's [or Richman's] Ifland is on the coaft of Maine, between Cape Elizabeth and Black Point. Joffelyn vifited it, in September, 1638; "where Mr. Tralanie [Trelawney] kept a fifhing. Mr. John Winter, a grave and difcreet man was his agent, and imployer of 60 men upon that defign." *Voyages to N. E.*, 25, 26. Winthrop (i. 124) mentions the coming of feventeen fifhing fhips to Richman's Ifland and the Ifles of Shoals, in the winter of 1633-4.

[185] June 21, 1641, Lechford drafted a "commiffion to John Webbe als [Evered] of Bofton and his company to trade and doe their bufineffe at the Ifle of Sables, and to paffe in the barke Endevor of Salem, whereof is mafter Jofeph Grafton." *Ms. Journal*, 224. "This fummer [1641] the merchants of Bofton fet out a veffel again to the Ifle of Sable, with 12 men, to ftay there a year. They fent again in the 8th month, and in three weeks the veffel returned" with 400 pair of fea-horfe teeth, worth £300. *Winthrop*, ii. 34, 35. Earlier expeditions, in 1635, 1637, and 1638-9, had been lefs fuccefsful. *Ibid.*, i. 162, 237, 305.

Vineyard,[186] uninhabited by any Englifh, but Indians, which are very favage.

46
French and
Dutch.

Northward from the *Bay*, or Northeaft, lyeth the *French* plantation, who take up bever there, and keepe ftrict government, boarding all veffels that come neare them, and binding the mafters till the governour, who is a Noble-man,[187] know what they are; and fouth of *New-England* the *Dutch* take up the bever.

Joffelyn mentions "the *Amphibious* creature, the *Walrus, Mors,* or *Sea-Horfe*,"—"a kind of monftrous-fifh numerous about the Ifle of *Sables,* i. e. The fandy Ifle." *Voyages,* 10, 106.

[186] "The Ifle of *Capawack.*" Bradford, 97. "Thofe of the Ifles of Capawack" fent to make friendfhip." *Ibid.,* 104. "The Ifle *Capewak* ... now called *Martin's Vineyard.*" Morton's *Memorial* (1669), 26. Winthrop wrote "Martin's Vineyard," when noticing the beginning of a plantation there by "fome of Watertown," in 1643 (ii. 151, 152). So, Thomas Mayhew himfelf, in 1650; Henry Whitfield ("*Martin's* Vineyard, ... fome call it *Marthaes* Vineyard"), in 1651; and Hubbard, a generation later. But "Martha's Vineyard" was the name given by Gofnold, in 1602, to the fmall ifland now called No-man's Land (3 *Mafs. Hift. Coll.,* viii. 75, 76); and the "Iflands of Capawock alias Martha's Vineyard" were, by that name, conveyed to Thomas Mayhew, Oct. 25,

1641. Hough's *Nantucket Papers,* 4. See 2 *Mafs. Hift. Coll.,* iv. 107, 118, 184; Belknap's *Amer. Biog.,* ii. 113; Davis's *Morton,* 58, 275. By Indians of the main land, the ifland was called *Nope.* 2 *Mafs. Hift. Coll.,* ii. 242.

[187] Charles d'Aulnay de Charnifé, governor of the divifion of Acadie which was weft of the river St. Croix. After the death (in 1635) of Razilly, chief commander of French Acadie, D'Aulnay and the Sieur de La Tour (to whom had been affigned the government of the eaftern divifion), quarrelled for the fucceffion, the former holding fortified pofts at Penobfcot (whence he had expelled the Plymouth traders in 1635), at Port Royal (now Annapolis) and La Hève (now New Dublin), in Nova Scotia. La Tour had a fort at the mouth of the St. John. See *Winthrop,* i. 117, 166, 171, 206; ii. 42, 43, 107-14, &c.; *Hutchinfon,* i. 127-135, 497-516; 3 *Mafs. Hift. Coll.,* vii. 90-121; and Palfrey's *Hift. of N. Eng.,* ii. 144-151.

Three hundred Leagues fouth from the *Bay* along the *Virginia.*
coafts, lyeth *Virginia;* neare to that is *Maryland,* where *Maryland.*
they are Roman Catholiques, they fay.
There was a fpeech of fome *Swedes* which came to *Swedes.*
inhabit neere *Delawar Bay,* but the number or certainty
I know not.
Three hundred leagues from the *Bay,* Eaftward, lyeth *New-found-land.*
New-found-land, where is a maine trade for fifhing. Here
we touched comming homeward.[188]

Florida lyes betweene *Virginia* and the *Bay* of *Mexico,* *Florida.*
and had been a better Country for the *Englifh* to have
planted in, according to the opinion of fome, but it is fo
neere the *Spaniard,* that none muft undertake to plant
there, without good Forces.

For the ftate of the Country in the Bay *and thereabouts.* 47

THe Land is reafonable fruitfull, as I think; they State of the Countrey of
have cattle, and goats, and fwine good ftore, and *New-England.*
fome horfes, ftore of fifh and fowle, venifon, and * corne, *Wheat and Barley are thought
both *Englifh* and *Indian.* They are indifferently well not to be fo good as thofe grains in
able to fubfift for victuall. They are fetting on the man- *England;* but the Rye and Peafe
ufacture of linnen and cotton cloath,[199] and the fifhing are as good as the Englifh: the
Peafe have no
wormes at all.
Beanes alfo there
are very good.[190]

[188] "There being nó fhip which was to return right for England, they went to Newfoundland, intending to get a paffage from thence in the fifhing fleet." *Winthrop,* ii. 31.

[189] This marginal note is not in the M.H.S. MS.

[190] The General Court, May, 1640, "taking into ferious confideration the abfolute neceffity for the raifing the

trade,[191] and they are building of ships,[192] and have good store of barks, catches, lighters, shallops, and other ves-

manufacture of linen cloth," made an order for the promotion of this branch of industry in the several towns, as also, "for the spinning and weaving of cotton wool." *Mass. Rec.*, i. 294. At the October session, a bounty was granted of 3d. on the shilling on the value of all linen, woollen, and cotton cloth which should be made in the jurisdiction, of yarn spun or materials raised therein. *Ibid.*, 303. The next year, payment of this bounty was ordered to be made on 83½ yards of cloth, valued at a shilling per yard; but the people did not approve the action of the Court, and the order of the preceding year was, at the request of the deputies, repealed by the General Court, June, 1641. *Ibid.*, 316, 320. Connecticut, in February, 1641, ordered that hemp or flax should be planted by every family in the jurisdiction, that "we might, in time, have supply of linen cloth among ourselves." *Conn. Rec.*, i. 61, 64.

"Rowley, to their great commendation, exceeded all other towns," in the manufacture of cloth, as Winthrop (ii. 119, 120) records, under the year 1643. The settlers of that town were mostly from Yorkshire, and "were the first people that set upon making of cloth in this Western world, ... many of them having been clothiers in England."—Johnson's *W. W. Providence*, b. ii. ch. 11.

[191] See before, p. 45, note 182. "This year [1641] men followed the fishing so well, that there was about 300,000 dry fish sent to the market." —*Winthrop*, ii. 42.

In July, 1640, Lechford drew an agreement between Mr. Thomas Fowle of Boston and John Squire, Nicholas Squire, and Sampson Anger [Angier], all of Acomenticus, fishermen, for the purchase of as many "merchantable dry cod-fish" as they should take, cure, and deliver to him on board vessels at or near the Isle of Shoals, within twelve months thereafter; for which he was to pay fourteen shillings per kental.—*Ms. Journal*, 155.

"Some of the freemen and inhabitants of Hingham" petitioned the General Court in June, 1641, to be "instituted into a company" for establishing a fishing plantation at Nantasket, and for a grant to themselves, for that purpose, of "the said neck of land called Nantasket, from sea to sea, unto the head of Straits pond." *Ibid.*, 221. The court granted the land, and gave liberal encouragement to the enterprise; and, in 1644, the plantation, having become a town, with "twenty houses and a minister," was named Hull. *Mass. Rec.*, i. 320, 326; *Winthrop*, ii. 175.

[192] "The general fear of want of foreign commodities, now our money was

Newes from New-England.

fels. They have builded and planted to admiration for the time. There are good mafts and timber for fhipping, planks, and boards, clap-board,[193] pipe-ftaves, bever, and furres, and hope of fome mines.[194] There are Beares, Wolves, and Foxes, and many other wilde beafts, as the Moofe, a kind of Deere, as big as fome Oxen, and gone, and that things were like to go well in England, fet us on work to provide fhipping of our own, for which end Mr. [Hugh] Peter, being a man of very public fpirit and fingular activity for all occafions, procured fome to join for building a fhip at Salem, of 300 tons, and the inhabitants of Bofton, ftirred up by his example, fet upon the building another at Bofton, of 150 tons." *Winthrop*, ii. 24, under date of Feb. 2, 1641. Both fhips were finifhed in 1641. *Ibid*, 31. Mr. Peters and Emanuel Downing write from Salem, Jan. 13, 1641, that there were "two or three fhips building" there. 4 *Mafs. Hift. Coll.*, vi. 90. The next year (1642), "five fhips more were built, three at Bofton, one at Dorchefter, and one at Salem" (*Winthrop*, ii. 65); and in September, the author of "New England's Firft Fruits" wrote (p. 22): "Befides many boats, fhallops, hoys, lighters, pinnaces, we are in a way of building fhips of an 100, 200, 300, 400 tons. Five of them are already at fea, many more of them in hand at this prefent," &c.

[193] If it were not for the perfiftent omiffion in modern dictionaries of the primary meaning of this word, it would be unneceffary to remark here, that it was applied to all fmall boards (efpecially to *paling* and *pipe-ftaves*) which were made by riving or *cleaving*, in diftinction from *fawed* boards. *Cloven* (A. Sax. *clough*) boards eafily paffed into "clo'-boards," "claw-boards," "clobboards," and "clap-boards." Joffelyn wrote of the "cleaving of clawboard," and of oak wood "excellent for claw-board and pipe-ftaves." *Voy.*, 208; *N. E. Rar.*, 48. Wood diftinguifhes between oaks "more fit for *clappboard*, [and] others for *fawne* board." *N. E. Profpect*, pt. i. c. 5.

[194] Comp. Joffelyn, *N. E. Rar.*, 92, 93; *Voyages*, 44; Wood's *N.E. Profpect*, pt. i. c. 5. John Winthrop, Jr., failed for England in the fame fhip with Lechford, and, while abroad, formed a company for eftablifhing an iron-work in New England; returning, in 1643, with £1000 ftock, and a number of workmen. See *Winthrop*, ii. 212, and Savage's note; *Mafs. Rec.*, i. 206, 327; ii. 61, 81, 125; 4 *Mafs. Hift. Coll.*, vi. 516, 517.

Lyons,[195] as I have heard. The Wolves and Foxes are a great annoyance. There are Rattle ſnakes, which ſometimes doe ſome harme, not much; He that is ſtung with any of them, or bitten, he turnes of the colour of the Snake, all over his body, blew, white, and greene ſpotted; and ſwelling, dyes, unleſſe he timely get ſome Snake-weed;[196] which if he eate, and rub on the wound, he may

[195] Everybody in New England had *heard* of theſe lions. "For beaſts, there are ſome bears, and they ſay ſome lions alſo; for they have been ſeen at Cape Anne. ... I have ſeen the ſkins of all theſe beaſts ſince I came to this Plantation, excepting lions." Higginſon's *N. E. Plantation* [in Young's *Chron. of Maſs.*, 248]. Wood, too, heard "ſome affirme that they have ſeene a Lyon at Cape Anne," and ſays that ſome who were loſt in the woods had "heard ſuch terrible roarings, as ... muſt eyther be *Devills* or *Lyons*. ... Beſides, Plimouth men have traded for Lyons ſkinnes in former times." *N. E. Proſpect*, pt. 1, ch. vi. Joſſelyn was told, at Black Point, of "a young Lyon (not long before) kill'd at Piſcataway by an Indian" (*Voyages*, 23); and there were ſome "yet living in the country," in 1663, or later, to affirm that a young lion had been ſhot by an Indian, not far from Cape Ann. *N. E. Rar.*, 21, 22. The ſuppoſed lion may have been the cougar, or puma, ſometimes called the American lion, or panther.

[196] "The Antidote to expell the poyſon ... is a root called ſnakeweed, which muſt be champed, the ſpittle ſwallowed, and the root applyed to the ſore; this is preſent cure againſt that which would be preſent death without it: this weed is ranck poyſon, if it be taken by any man that is not bitten. ... Cowes have been bitten, but being cut in divers places, and this weede thruſt into their fleſh were cured." Wood's *N. E. Proſpect*, pt. i. ch. xi. Higginſon (*N. Eng. Plantation*) ſays, the "ſting" of the rattle-snake will cauſe death "within a quarter of an hour after, except the party ſtinged have about him ſome of the root of an herb called ſnake-weed to bite on, and then he ſhall receive no harm." Young's *Chron. of Maſs.*, 255. Cornuti (*Canadenſium Plantarum*, &c. Paris, 1635), as cited by Prof. Tuckerman in his Introduction to Joſſelyn's *N. E. Rarities*, mentions a root received *ex notha Anglia*, "known, it appears, by the name of *Serpentaria*, or, in the vernacular, *Snaqrocl*,—a ſure remedy for the bite of a huge

haply recover, but feele it a long while in his bones and body. Money is wanting, by reafon of the failing of paffengers thefe two laft yeares, in a manner. They want help to goe | forward, for their fubfiftence in regard of cloathing: And great pity it would be, but men of eftates fhould help them forward. It may bee, I hope, a charitable worke. The price of their cattell, and other things being fallen,[97] they are not at prefent able to make fuch returns to *England*, as were to be wifhed for them: God above direct and provide for them. There are multitudes of godly men among them, and many poore ignorant foules. Of late fome thirty perfons went in two small Barks for the *Lords Ifle of Providence*,[198] and for the

48

and moft pernicious ferpent." Prof. T. thinks this to be "one of the numerous varieties of *Nabalus albus* (L.) Hook., if not, as Purfh fuppofed, what is now the *var. Serpentaria*, Gray." *Trans. Amer. Antiq. Soc.*, iv. 119. Joffelyn figures and defcribes the *Nabalus albus*, in *N. E. Rarities*, 76, but without allufion to its virtues.

Gov. Winthrop mentions (i. 62) that "he always carried about him ... in fummer time, snakeweed."

[97] See *Winthrop*, ii. 7, 18, 21, 24; *Mafs. Col. Rec.*, i. 304. 307; E. Winflow's letter from Plymouth, June, 1640, in 4 *Mafs. Hift. Coll.*, vi. 166. In the fummer of 1641, "few coming to us, all foreign commodities grew fcarce, and our own of no price. Corn would buy nothing;" a cow worth £20 in 1640 might now be bought for £4 or £5: "fo as no man could pay his debts, nor the merchants make return into England for their commodities." *Winthrop*, ii. 21, 31.

[198] Lechford left New England before the return of thefe barks, with their paffengers (Sept. 3, 1641), made known the difaftrous iffue of this expedition, "which brought fome to fee their error, and acknowledge it in the open congregation, but others *were* hardened." *Winthrop*, ii. 33, 34. The provifions of the charter granted in 1630 to the Adventurers for the Plantations of the Iflands of Providence, Henrietta, and the adjacent iflands (the Bahamas), were very liberal, and

Maine thereabout, which is held to be a beter countrey and climate by some: For this being in about 46. degrees of northerne latitude, yet is very cold in winter, so that some are frozen to death, or lose their fingers or toes every yeere, sometimes by carelesnes, sometimes by accidents, and are lost in snowes, which there are very deepe sometimes, and lye long: Winter begins in October, and lasts till Aprill.[199] Sixty leagues Northerly it is held not habitable, yet again in Summer it is exceeding hot. If shipping for conveyance were sent thither, they might spare divers hundreds of men for any good design.[200] The jurisdiction of the *Bay* Patent reacheth from *Pascattaqua* Patent Northeast to *Plymouth* Patent Southward. And in my travailes there, I have seene the towns of *Newberry*, *Ipswich*, *Salem*, *Lynne*, *Boston*, *Charlestowne*, *Cambridge*, *Watertowne*, *Concord*, *Roxbury*, *Dorchester*, and *Braintree* in the *Bay* Patent, *New Taunton* in *Plym*-

the Company offered great encouragement to planters. "The great advantages supposed to be had in Virginia and the West Indies, &c., made this country to be disesteemed of many," wrote Winthrop, in 1640. John Humfrey, appointed by the Company in February, 1641, Governor of Providence Island, "labored much to draw men to join with him." But, before the emigrants from New England arrived at Providence, the island had been taken by the Spaniards, who fired upon one of the vessels when coming into harbor, and within pistol-shot of the fort, and killing her commander, William Peirce, and Mr. Samuel Wakeman of Connecticut, a passenger. *Winthrop*, i. 332, ii. 33, 34; Johnson, *W. W. Providence*, b. 2, ch. 20.

[199] For "Aprill," the M.H.S. MS. has "March."

[200] The section ends here in the M.H.S. MS. The eleven lines which follow were subsequently added.

outh Patent, the Ifland *Aquedney*, and the two townes therein, | *Newport* and *Portfmouth*, and *New Providence* within the *Bay* of *Narhigganfets*. This for the fatisfaction of fome that have reported I was no Travailer in *New-England*.

49

Concerning the Indians, *or Natives.*

They are of body tall, proper, and ftraight; they goe naked, faving about their middle, fomewhat to cover fhame. Seldome they are abroad in extremity of Winter, but keep in their *wigwams*, till neceffity drives them forth; and then they wrap themfelves in fkins, or fome of our Englifh coorfe cloth: and for the Winter they have boots, or a kind of laced tawed-leather ftockins. They are naturally proud, and idle, given much to finging, dancing, and playes; they are governed by *Sachems*, Kings; and *Saggamores*, petie Lords;[201] by an abfolute tyrannie. Their women are of comely feature, induftrious, and doe moft of the labour in planting, and carrying

Of the *Indians*.

---

[201] This diftinction is not well founded. *Sachem* and *Sagamore* were two forms of the fame word, —*fagkiman*, "he leads," "directs." Wood's vocabulary has, "*Sagamore*, a king. *Sachem*, idem." Dudley. in his letter to the Countefs of Lincoln, writes, that "Chickatalbott . . . leaft favoreth the Englifh, of any fagamore (for fo are the kings with us called, as they are *fachims*, fouthwards)," &c. Young's *Chron. of Mafs.*, 305. Capt. John Smith makes a fimilar diftinction : "The *Maffachufets* call . . their Kings *fachemes*. The *Pennobfcots*, . . *fagamos.*" *Advert. for the Unexper.*, 3 *Mafs. Hift. Coll.*, iii. 23. Comp. Joffelyn's *Voyages*, 123.

of burdens; their husbands hold them in great flavery, yet never knowing other, it is the leffe grievous to them. They fay, *Englifhman* much foole, for fpoiling good working creatures, meaning women: And when they fee any of our *Englifh* women fewing with their needles, or working coifes, or fuch things, they will cry out, Lazie *fquaes!* but they are much the kinder to their wives, by the example of the *Englifh.* Their children, they will not part with upon any terms, to be taught. They are of complexion fwarthy and tawny; | their children are borne white, but they bedawbe them with oyle, and colours, prefently. They have all black haire, that 1 faw.

In times of mourning, they paint their faces with black lead, black, all about the eye-brows, and part of their cheeks. In time of rejoycing, they paint red, with a kind of vermilion. They cut their haire of divers formes, according to their Nation or people, fo that you may know a people by their cut; and ever they have a long lock on one fide of their heads, and weare feathers of Peacocks, and fuch like, and red cloath, or ribbands at their locks; beads of *wampompeag* about their necks, and a girdle of the fame, wrought with blew and white *wampom*, after the manner of chequer work, two fingers broad, about their loynes: Some of their chiefe men goe fo, and pendants of *wampom*, and fuch toyes in their ears. And their women, fome of the chiefe, have faire bracelets, and chaines of *wampom*. Men and women, of them, come confidently among the *Englifh*. Since the *Pequid* war, they are kept in very good fubjeﬆion, and held to ﬆriﬆ points of Juﬆice, fo that the *Englifh* may travail fafely among them. But the *French* in the Eaﬆ, and the *Dutch* in the South, fell them guns, powder and fhot.[202] They have *Powahes*, or Prieﬆs, which are Witches, and a kind of Chirurgions, but fome of them, notwithﬆanding, are faine to be beholding to the *Englifh* Chirurgions. They will have their times of *powaheing*, which they will, of late, have called Prayers, according to the *Englifh* word. The | *Powahe*[203] labours himfelfe in his incantations, to

[202] De Vries, in an account of the Indians of New Netherland, in 1640, fays, "They have now obtained guns from our people [the Dutch]. He was a villain who firﬆ fold them to them, and fhowed them how to ufe them." *Voyages* (tranflated by Murphy), in 2 *N. Y. Hiﬆ. Soc. Coll.*, iii. 95. Comp. Brodhead's *New York*, i. 308 ; *Records of Comm'rs of U. Cols.* (Hazard, ii.) 19, 58. Bradford (*Hiﬆ. of Plymouth*, 238, 337) complains of the French trade in arms and ammunition ; but, in another place, he diﬆributes more impartially his cenfure of the "bafenefs of fundry unworthy perfons, both *Englifh*, *Dutch*, and *French*," who had fupplied the Indians of thefe parts with "peeces, powder and fhote" (pp. 235, 238–9).

[203] *Powwáw*, as Roger Williams writes it. *Pauwau*, Eliot. This word is nearly related to, if not identical with, *taúpowaw*, "a wife fpeaker ; " pl. *taupowaûog*, " their wife men, and old men (of which number their Prieﬆs are alfo)." R. Williams, *Key*, 57, 120. Wood (*N. E. Profpeﬆ*, pt. 2, c. xii.) gives an amufing account of the "pow-wows" and their doings. He admits "that, by God's permiffion, through the Devils helpe, their charms are of force to produce effeﬆs of wonderment," and fays, "fometimes the Devill for requitall of their worfhip, recovers the partie [who is fick or lame] to nuzzle them up in their devillifh Religion." Comp. R. Williams, *Key*, c. xxi.; Winflow's *Good Newes from N. E.* [2 *Mafs. Hiﬆ. Coll.*, ix. 92, 93].

extreame fweating and wearineffe, even to extafie. The *Powahes* cannot work their witchcrafts, if any of the *Englifh* be by; neither can any of their incantations lay hold on, or doe any harme to the *Englifh*, as I have been credibly informed. The *Powahe* is next the King, or *Sachem*, and commonly when he dyes, the *Powahe* marryes the *Squa Sachem*, that is, the queene. They have marriages among them; they have many wives; they fay, they commit much filthineffe among themfelves. But for every marriage, the *Saggamore* hath a fadome of *wampom*, which is about feven or eight fhillings value. Some of them will diligently attend to any thing they can underftand by any of our Religion, and are very willing to teach their language to any *Englifh*. They live much the better, and peaceably, for the *Englifh;* and themfelves know it, or at leaft their *Sachems*, and *Saggamores* know fo much, for before they did nothing but fpoile and deftroy one another.[204] They live in *Wigwams*, or houfes made of mats like little hutts, the fire in the midft of the

[204] "The Pagan world of Indians here will acknowledge our fitting down by them, hath prevented the danger either of their diffolution or fervitude. For the Indians in thefe parts being by the hand of God fwept away (many multitudes of them) by the plague, the manner of the Neighbor-Indians is either to deftroy the weaker Countreys, or to make them Tributary: which danger ready to fall upon their heads in thefe parts, the coming of the Englifh hither prevented." Cotton's *Way of Congr. Churches cleared*, pt. i. p. 21. See alfo Higginfon's *N. E. Plantation*, in Young's *Chron. of Mafs.*, 257; Wood's *N. E. Profpect*, pt. i. ch. 9.

houſe. They cut downe a tree with axes and hatchets, bought of the *Engliſh*, *Dutch*, or *French*, & bring in the butt-end into the *wigwam*, upon the hearth, and ſo burne it by degrees. They live upon parched corne,[205] (of late, they grinde at our *Engliſh* mills.) Veniſon, Bevers, Otters, Oyſters, Clammes, Lobſters, and other fiſh, Ground-nuts,[206] Akornes, they boyle all together in a kettle. Their riches are their *wampom*, bolles, trayes, | kettles, and ſpoones, bever, furres, and canoos. He is a *Sachem*, whoſe wife hath her cleane ſpoons in a cheſt, for ſome chief *Engliſh* men, when they come on gueſt wiſe to the *wigwam*. They lye upon a mat, with a ſtone, or a piece of wood under their heads; they will give the beſt entertainment they can make to any *Engliſh* comming amongſt them. They will not taſte ſweet things, nor alter their habit willingly; onely they are taken with tobacco, wine, and ſtrong waters; and I have ſeene ſome of them in

[205] "*Nŭkchick*, parch'd meale ... which they eate with a little water, hot or cold." R. Williams, *Key*, p. 11. (ch. ii.) "*Nocake* (as they call it) which is nothing but Indian corne parched in the hot aſhes," and "afterwards beaten to powder." Wood, *N. E. Proſpect*, pt. 2, ch. vi.

[206] "Earth-nuts, which are of divers kinds, — one bearing very beautiful flowers," (which Prof. Tuckerman identifies with the *Apios tubero-ſa*, Moench.) Joſſelyn's *N. E. Rar.*, 47 (*Trans. A. A. Soc.*, iv. 180).— Brereton noted the "great ſtore of ground-nuts" to be found "in every iſland, and almoſt in every part of every iſland;" "forty together on a ſtring, ſome of them as big as hen's eggs; they grow not two inches under ground: the which nuts we found to be as good as potatoes." *Account of Goſnold's Voyage*, 3 *Maſs. Hiſt. Coll.*, viii. 89.

English, or *French* cloathes. Their ordinary weapons are bowes and arrowes, and long ſtaves, or halfe pykes, with pieces of ſwords, daggers, or knives in the ends of them: They have Captaines, and are very good at a ſhort mark, and nimble of foot to run away. Their manner of fighting is, moſt commonly, all in one fyle. They are many in number, and worſhip *Kitan*,[207] their good

[207] Comp. E. Winſlow's *Good Newes from N. E.* (Young's *Chron. of the Pilgr. Fathers*, 326, 355): Wood's *N. E. Proſpect*, pt. ii. ch. 12. "The Maſſachuſets call their great God *Kichtan*, . . . and that we ſuppoſe their Devill, they call *Habamouk*. The Pennobſcots, their God, *Tantum*." J. Smith's *Advert. for the Unexperienced*, ch. vi.

Higginſon (in *N. E. Plantation*), wrote: "For their religion, they do worſhip two Gods, a good God and an evil God. The good God they call *Tantum*, and their evil God, whom they fear will do them hurt, they call *Squantum*." Robert Southey, tranſcribing this "very ſummary account" of the Indian faith, adds: "An equal degree of knowledge on the part of the Indians might have made them deſcribe Mr. Higginſon himſelf as a *Squantumite*." Southey's *Com.-Place Book*, 2d Ser., 656. The comment, though miſchievous, is not wholly unjuſt. Had our early writers been more diligent ſtudents of the Indian language, they would have diſcovered, probably, that *Tantum* and *Squantum* were names of the ſame "Great Spirit," or *Keihtan*, — to be worſhipped as a beneficent, or propitiated as an angry, god. *Squantum*, or *m'ſquantum*, ſignifies, "he is angry" [*lit.*, bloody-minded]. "If it be but an ordinary accident, a fall, &c., they will ſay, *God was angry* and did it. *Muſquàntumìnanit*, God is angry." R. Williams, *Key*, p. 115.

Manit, the word which is often tranſlated "God," conveyed to the Indian no other or higher idea than that of ſomething *extra-ordinary* and tranſcending former experience. Its literal ſignification is, "that which ſurpaſſes," "that which is *more than*," other perſons or things with which it is compared. "At the apprehenſion of any Excellency in Men, Women, Birds, Beaſts, Fiſh, &c. [they] cry out *Manittóo*, that is, it is a God;" and this "they ſay of every thing which they cannot comprehend." R. Williams, *Key*, 118, 105. The initial *m* repreſents the imperſonal prefix, while *anit* is a regularly-formed verbal.

Newes from New-England.

god, or *Hobbamocco*,[208] their evil god; but more feare *Hobbamocco*, becaufe he doth them moft harme. Some of their Kings names are *Canonicus, Meantinomy*,[209] *Owſhamequin*,[210] *Cuſhamequin*,[211] *Webbacowitts*, and *Squa Sachem*,[212] his wife: She is the Queene, and he is *Powahe*,

From *keihte*, 'great,' 'chief,' and *anit*, is formed *keihtannit*, "great fuperior being" [which Eliot ufed in tranflating Genefis xxiv. 7, "the LORD God," *Jehovah Keihtannit*.] Of this word, *Kiehtan, Kitan*, &c., were contract forms, or equivalents. Comp. the Narraganfet *Kautántowit*, "the great South-weft God" (R. Williams, *Key*, 116); the Delaware *Getaunitowit* (Heckew.); and the Old Algonkin *kitchi manitoo* (Lahontan).

[208] "*Hobbamock*, as they call the Devil." *Winthrop*, i. 254. "*Abamocho* (the Devill) whom they much feare." Wood's *N. E. Proſpect*, pt. 2, ch. viii. "*Abbamocho* or *Cheepie*." Joffelyn's *Voyages*, 132.

[209] *Caunoũnicus*, and his nephew, *Miantunnômu*, fachems of the Narraganfetts.

[210] *Oufamequin, Ofomeagen, Ofamekin, Afuhmequin*, &c., as the name is varioufly written; the great fachem of the Wampanoags, — better known as *Maſſaſoit*. His principal refidence was at Sowams, now Warren, R. I. See Dexter's *Mourt's Relation*, 91, 94, 98, &c.; *Bradford*, 94, 102, &c.

[211] *Cutſhamakin* was the nominal chief of the few remaining Indians of Neponfet. Chickataubut, who lived "upon the river of Naponfet, near to the Maffachufetts Fields," (in Quincy,) was "the greateft fagamore in the country" (as Wood was told,) before the plague of 1616-18 fwept over this part of New England. In 1631, he had only between fifty and fixty fubjects; and many of thefe, with the fachem himfelf, died of fmall-pox in 1633. "Jofias, Chickatabot his heir" was not then of age, and Cutfhamakin, who is faid to have been a brother of Chickataubut, and who had been a humble hanger-on of the Englifh from their firft coming, fucceeded for a time to the titulary honor of fachem of Maffachufetts, and to the right of figning deeds and conveyances of lands once occupied by the tribe. *Winthrop*, i. 48, 116, 192, 195, &c., ii. 153; Wood's *N. E. Proſpect*, pt. 1, c. x.; Dudley's *Letter*, in Young's *Chron. of Maſs.*, 305; *Hiſt. of Dorcheſter*, 10, 11, 47; Gookin, 1 *Maſs. Hiſt. Coll.*, i. 169.

[212] "*Webcowites*, and the Squa Sachem of Mifticke, wife of the faid Webcowites." Lechford's *Ms. Journal*, 143. "Squa Sachem & *Webba Cowet*." *Maſs. Rec.*, i. 201. The Squa Sachem had been the wife of

and King, in right of his wife. Among some of these Nations, their policie is to have two Kings at a time; but, I thinke, of one family; the one aged for counsell, the other younger for action. Their Kings succeed by inheritance.

M. Dunster a hopefull Schoolmaster.

53

Master *Henry Dunster*, Schoolmaster of *Cambridge*, deserves commendations above many; he | hath the plat-forme and way of conversion of the Natives, indifferent right, and much studies the same, wherein yet he wants not opposition, as some other also have met with: He will, without doubt, prove an instrument of much good in the Countrey, being a good Scholar, and having skil in the Tongues; He will make it good, that the way to instruct the *Indians*, must be in their *owne* language, not *English*;[213] and that their language may be perfected.[214]

Nanepashemit, the great sagamore of the Pawtucket Indians (north and east of Charles River), who was killed by the Tarratines in 1619. His sons, Wonohaquaham, or Sagamore John, of Mistick, "the chiefest Sachim in these parts, at our first coming hither" (Cotton's *Way cleared*, i. 80), and Montowompate, or Sagamore James, of Saugus, with most of their people, died of small-pox in Dec., 1633. The widow married Webbacowit before 1635. One of her places of residence is supposed to have been in what is now West Cambridge. In 1644, she, with other petty sachems, made a formal submission to the government of Massachusetts. See *Winthrop*, i. 119; Dexter's *Mourt's Relation*, 126–28 ; Brooks's *Medford*, 73, 74; Young's *Chron. of Mass.*, 306, 307 ; Frothingham's *Hist. of Charlestown*, 32–36.

[213] See Mr. Dunster's letter to Dr. Ravis, in 4 *Mass. Hist. Coll.*, i. 251-54. He writes: "We do not trouble the Indians to learn our English, but onely such as for their owne behoof doe it of their owne accord."

[214] Near the end of this paragraph, Lechford, in his earlier draft, had in-

Newes from New-England.

A Note of some late occurrences touching Epiſcopacie.

Some of the learnedſt, and godlieſt in the *Bay*, begin to underſtand Governments; that it is neceſſary, when Miniſters or People fall out, to ſend other Miniſters,

Some late occurrences concerning Epiſcopacie.

ſerted: "Mrˢ Glover did worthily and wifely to marry him." *M.H.S. MS.* Mrs. Elizabeth Glover, who married Mr. Dunſter in June, 1641, was the widow of Rev. Joſſe Glover, rector of Sutton, co. Surrey, from 1628, or earlier, till December, 1634, when he was ſuſpended for refuſing to read the book of ſports. He was "much beloved of moſt, if not of all, and his departure lamented by moſt, if not of all," his people, as the pariſh regiſter affirms. (His ſucceſſor was inducted June 10, 1636.) He is afterwards deſcribed as "of London"; but his reſidence there muſt have been brief, for he ſailed for New England in 1638, with the intention of eſtabliſhing a printing preſs here, having made a contract, June 7, 1638, with Stephen Day of Cambridge to come over for that purpoſe.

Mr. Glover died on the paſſage. His will, which was probably executed before ſailing, names the Rev. John Harris, D.D., Warden of Wincheſter College, and Richard Davys, merchant, of London, his executors. He left two ſons, *Roger* and *John* (H. C., 1650), and three daughters, *Elizabeth* (who married Adam Winthrop), *Sarah* (who married Deane Winthrop), and *Priſcilla* (who married John Appleton). There may have been other children whoſe names do not appear. Of theſe five, Roger, Elizabeth, and Sarah, were by a former wife, Sarah, (daughter of Roger Owfield of London,) who died at Sutton, July 10, 1628, aged 30 years, while her huſband was rector there. Her epitaph, with the names of her children, may be ſeen in Manning & Bray's *Hiſt. of Surrey*, ii. 483.

Mr. Glover's name frequently occurs in Lechford's Ms. Journal, variouſly written *Joas, Jofs,* and *Joſſe* Glover. On his wife's monument, and in the extract from the pariſh regiſter of Sutton, it is *Joſeph;* and elſewhere it appears as *Jeſſe. Winthrop,* i. 289; Manning & Bray's *Hiſt. Sur.,* ii. 487; Lechford's *Ms. Journal; Calendar of (Brit.) St. Papers, Dom. Ser.,* 1634-5, p. 355; Savage's *Geneal. Dict.;* Thomas's *Hiſt. of Printing,* i. 222-26, 458-66. From Mr. Dunſter's ſtatement of account with the eſtate (printed by Thomas, from the County Court Records), it appears that his

or they voluntarily to goe among them, to feek by all good wayes and meanes to appeafe them.²¹⁵

And particularly, Mafter *Peter* went from *Salem* on foot to *New Dover*, alias *Pafcattaqua*, alias *Northam*, to appeafe the difference betweene Mafter *Larkham* and Mafter *K.* when they had been up in Armes this laft Winter time.²¹⁶ He went by the fending of the *Governour*, *Counfell*, and *Affiftants* of the *Bay*, and of the Church of *Salem;* and was in much danger of being loft, returning, by lofing his way in the woods, and fome with him, but God be bleffed they returned.

---

wife died "two years and two months after her marriage" with him, — that is, about Auguft, 1643.

²¹⁵ "Who giving advice according to the Word, doe by the bleffings of Chrift heale jealoufies, and compofe differences, and fettle peace and love amongft them." Cotton, *Way of the Churches*, 106.

"When a Congregation wanteth agreement and peace amongft themfelves, it is then a way of God (according to the patterne, Acts 15. 2.) to confult with fome other Church, or Churches, either by themfelves or their meffengers met in a Synod. But then they fend not to them for power to adminifter any ordinance amongft themfelves, but for light to fatisfie diffenters, and fo to remove the ftumbling-block of the fufpition of maladminiftration of their power, out of the way." Cotton, *Way cleared*, pt. i. pp. 94–5.

²¹⁶ See before, page 44, and *Winthrop*, ii. 28, 29. Hugh Peters's miffion to Pifcataqua had lefs to do with "epifcopacie" than Lechford fuppofed. "A good part of the inhabitants there" defired to come under the government of Maffachufetts; and this, as Winthrop believed, was the real caufe of the "eager profecution of Capt. Underhill" and his friends. It was on the petition of Underhill and the Maffachufetts party, for aid, that the governor and council gave commiffion, early in 1641, to "Mr. Bradftreet, one of our magiftrates, Mr. Peter and Mr. Dalton, two of our Elders, to go thither and to endeavor to reconcile them, and if they could not effect that, then to inquire how things ftood, and to certify

Againe he went a fecond time, for appeafing | the 54
fame difference, and had a Commiffion to divers Gentlemen, mafter *Humphrey*, mafter *Bradftreate*, Captaine
Wiggon, and mafter *Simons*, to affift, and to heare and
determine all caufes civill and criminall, from the *Governour of the Bay*, under his hand,²¹⁷ and the publique
feale, and then mafter *K.* went by the worft.

Mafter *Wilfon* did lately ride to *Greens harbour*,²¹⁸ in
Plymouth Patent, to appeafe a broyle betweene one mafter *Thomas*, as I take it, his name is, and mafter *Blindman*,²¹⁹ where mafter *Blindman* went by the worft, and

us, *etc.*" (*Winthrop*, ii. 28.) Mr. Peters, in a letter to Winthrop (without date, but which appears to have been written in the fpring of 1641), makes brief report of his miffion: "They there are ripe for our Gouernment as will appeare by the note I have fent you. They grone for Gouernment and Gofpell all ouer that fide on the Country. I conceive that 2 or 3 fit men fent ouer may doe much good at this confluxe of things ... If Mr. Larkham fay and hold, hee hath promifed mee to clofe with us," &c. 4 *Mafs. Hift. Coll.*, vi. 106-7. Not long afterwards (June 14, 1641), the proprietors of the Dover and Strawberry-Bank patents made a formal furrender of their jurifdiction to Maffachufetts; "whereupon a commiffion was granted to Mr. Bradftreet and

Mr. Simonds, with two or three of Pafcataquack, to call a court there and affemble the people to take their fubmiffion, etc., but Mr. Humfrey, *Mr. Peter and Mr. Dalton had been fent before* to underftand the minds of the people, to reconcile fome differences between them, and *to prepare them.*" *Winthrop*, ii. 38, 42 ; comp. *Mafs. Col. Rec.*, i. 324, 332.

²¹⁷ The original draft of this commiffion, dated July 8th, is in Lechford's Journal. It names as commiffioners, "John Humfrey Efq., Simon Bradftreete gent., Thomas Wiggon gent., and [Samuel] Symmons gent."

²¹⁸ See before, p. 41, note 137.

²¹⁹ Of the occafion of difference between Mr. Blinman and Mr. William Thomas, I can learn nothing. The fact of diffention and feparation is

Captaine *Keayne* and others went with mafter *Wilfon* on horfeback.

Alfo at another time, mafter *Wilfon*, mafter *Mather*, and fome others, going to the ordination of mafter *Hooke* and mafter *Streate*, to give them the right hand of fellow-fhip, at *New Taunton*, there heard the difference be-tweene mafter *Hooke* and mafter *Doughty*, where mafter *Doughty* was overruled, and the matter carried fomewhat partially, as is reported.[220]

It may be, it will be faid, they did thefe things by way of love, and friendly advife: Grant that; But were not the counfelled bound to receive good counfell? If they would not receive it, was not the Magiftrate ready to

briefly mentioned in the Plymouth Church record (i. 36), for the follow-ing extract from which, I am indebt-ed to my friend, the Rev. Henry M. Dexter, D.D.: —

"This church of Marfhfield, above called Green's Harbour, was begun, and afterwards carried on by the help and affiftance, under God, of Mr. Ed-ward Winflow, who att the firft pro-cured feverall Welfh Gentlemen of good note thither, with Mr. Blinman, a Godly able minifter, who unani-moufly joined together in holy fellow-fhip, or at leaft were in a likely way thereunto, but fome diffenfions fell amongft them, which caufed a parting not long after, and foe the hope of a Godly fociety as to them was fruf-trated. Not long after thofe that went from Plymouth with that Godly gentleman Mr. William Thomas, keeping up a communion, it pleafed the Lord to fend unto them a faithful and able preacher of the Gofpel, namely Mr. Edward Buckley, who was chofen their paftor, and officiated in that place very profitably divers years," &c.

Rev. Ebenezer Alden, Jr., in his *Sketch of the Church in Marfhfield*, p. 3, fays: "In confequence of a want of harmony between the new and old fettlers, after a refidence of a few months, Mr. Blinman, with moft of his friends, removed to Gloucefter." Comp. *Bradford*, 303-4; *Winth.*, ii. 64.

[220] See before, p. 41, note 136.

aſſiſt, and in a manner ready, according to duty, to *enforce* peace and obedience?[221] did not the Magiſtrates *aſſiſt*? and was not maſter *K.* ſent away, or compounded with, to ſeek a new place at *Long Iſland*,[222] maſter *Doughty* forced to the Iſland *Aquedney*,[223] and maſter *Blindman* to *Connecticot*?[224]

[221] How Mr. Cotton would have anſwered theſe queſtions may be inferred from ſome remarks of his on the relation of the Church to the civil magiſtrate, in a Thurſday lecture (preached early in 1640): "There is nothing more diſproportionable to us, then for us to affect Supremacy, for us to weare the hornes that might puſh Kings; to throw downe any, or *to deſire magiſtrates to execute what we ſhall think fit, verily it is not compatible to the ſimplicity of the Church of Chriſt.* Neither may they give their power to us, nor may we take it from them. . . . It is good to have theſe two States [the Church and the Magiſtracy] ſo joyned together, that the ſimplicity of the church may be maintained and upheld and ſtrengthened by the civill State according to God, but not by any ſimplicity further then according to the word. *Beware of all ſecular power*, and Lordly power, of ſuch vaſt inſpection of one church over another: Take heed of any ſuch uſurpation, it will amount to ſome monſtrous Beaſt: Leave every church Independent, *not Independent from brotherly counſell;* God forbid that we ſhould refuſe that; but when it comes to power, that one Church ſhall have power over the reſt, then look for a Beaſt [the alluſion is to Revel. xiii. 2], which the Lord would have all his people to abhor." — *Expoſition upon 13th chap. of Revelation*, pp. 30, 31.

[222] See before, p. 43, note 167.
[223] See before, pp. 40, 41, note 144.
[224] See note 219, above. Mr. Blinman removed, with his friends, to Cape Ann. Nothing is known of his going to *Connecticut* after leaving Marſhfield, before 1650, when he went from Glouceſter to New London, where he preached for ſeveral years. Perhaps Lechford was miſinformed as to the place of Mr. Blinman's new ſettlement; for, in a notice of Cape Ann (p. 45, ante), the coming of his company from Marſhfield is not mentioned. Poſſibly, however, Mr. B. did, in the firſt inſtance, direct his courſe from Plymouth Colony to Connecticut or New Haven, "to ſeek a new place," which not finding to his mind, or failing to ſecure ſatisfactory accommodations for himſelf and people, he returned eaſtward.

Questions to the Elders of Boston, *delivered* 9. *Septemb.*
1640.

1. WHether a people may gather themselves into a Church, without a Minister *sent* of God? [225]
2. Whether any People, or Congregation, may *ordaine* their owne Officers?
3. Whether the Ordination, by the hands of such as are *not Ministers*, be good? [226]

To the which I received an Answer the same day:

TO the first, the Answer is affirmative; for though the people in this Countrey are not wont to gather themselves into a Church, but (as you would have it) with the presence and advice of sundry Ministers; yet it were lawfull for them to gather into a Church without them. For if it be the priviledge of every Church to choose their owne Ministers, then there may be a Church, before they have Ministers of their owne; for Ministers of another Church have no power but in their owne Church.

[225] In a copy of these questions (in short-hand) in Lechford's Ms. Journal, the words "to approve thereof" are added at the end of the first question.

[226] In the manuscript (short-hand), the third question reads as follows: "Whether such as never had ordination or imposition of hands of the Presbyterie themselves, may warrantably impose hands upon any to the ministry? and if they do, whether it be good?"

Newes from New-England.

To the second and third; The second and third *Quef-tions* are coincident, and one Answer may serve for both: The Children of *Israel* did impose hands upon the *Levites, Num.* 8. 10. and if the people have power to elect their owne officers, they have power also to ordaine them; for Ordination is but an Installment of a man into that | office, whereto election giveth him right, nevertheleffe such a Church as hath a *Presbyterie*, ought to ordaine their Officers by a *Presbyterie*, according to 1 *Tim.* 4. 14. *This Answer was brought me by Master* Oliver, *one of the Elders, and Master* Pierce, *a Brother of* Boston.

When I was to come away, one of the chiefest[227] in the Country wished me to deliver him a note of what things I misliked in the Country, which I did, thus:

I doubt,

1. WHether so much time should be spent in the publique Ordinances, on the Sabbath day, becaufe that thereby some necessary duties of the Sabbath must needs be hindered, as visitation of the sick, and poore, and family.

2. Whether matters of offence should be publiquely handled, either before the whole Church, or strangers.

---

[227] This may have been the new governor, Richard Bellingham, with whom Lechford appears to have maintained very friendly intercourse.

3. Whether so much time should be spent in particular catechizing those that are admitted to the communion of the Church, either men or women; or that they should make long speeches; or when they come publiquely to be admitted, any should speak contradictorily, or in recommendation of any, unlesse before the Elders, upon just occasion.

4. Whether the censures of the Church should be ordered, in publique, before all the Church, or strangers, other then the denunciation of | the censures, and pronunciation of the solutions.

5. Whether any of our *Nation* that is not extremely ignorant or scandalous, should bee kept from the Communion, or his children from *Baptisme*.

6. That many thousands in this Countrey have forgotten the very principles of Religion, which they were daily taught in *England*, by set forms and Scriptures read, as the Psalmes, first and second Lesson, the ten Commandments, the Creeds, and publique catechizings. And although conceived Prayer[228] be good and holy, and so publike explications and applications of the Word, and also necessary both in and out of season: yet for the most part

---

[228] "In *conceived* prayer, the Spirit of God within us teacheth us what to pray. . . . But in *stinted* prayer, the matter is not suggested or endited to us by the Spirit of God within us, but prescribed and imposed upon us by the will, wisdom and authority of men," &c. — Cotton's *Answer to Mr. Ball's Discourse*, &c., ch. ii. [*Hanbury*, ii. 159.]

it may be feared they dull, amaze, confound, difcourage the weake and ignorant, (which are the moft of men) when they are in ordinary performed too tedioufly, or with the neglect of the Word read, and other premeditated formes inculcated, and may tend to more ignorance and inconvenience, then many good men are aware of.

7. I doubt there hath been, and is much neglect of endeavours, to teach, civilize, and convert the *Indian Nation*, that are about the Plantations.

8. Whether by the received principles, it bee *poſſible* to teach, civilize, or convert them, or when they are converted, to maintain Gods worſhip among them.

9. That electorie courfes will not long be fafe here, either in Church or Common-wealth.

10. That the civill government is not fo equally adminiſtred, nor can be, divers orders or by-laws confidered.

11. That unleffe thefe things be wifely and in time prevented, many of your ufefulleft men will remove and fcatter from you.

At *Boſton* July 5. 1641.

Certain Quæres about Church government, planting Churches, and some other Experiments.

a Mat. 10. 1.
Mar. 3. 13.
Act. 1. 4. & 2. 47.
& 3. 5. 6. & 9. 32.
35. & 11. 19. 20.
21. 26.
b Acts 8. 14. &
9. 31. & 11. 22. 27.
& 13. 2. 3. & 14.
21. to 28.
c Col. 2. 5.
Act. 11. 27. &
15. 36. & 16. 4.

1. WHether the people should cal the Minister, or the Minister ᵃ gather the people?

2. When a Church is gathered or planted; should they not have a care in ᵇ propagating other Churches, in other places next them.

3. Whether should not the first Church ᶜ visit the later Churches planted by them, to see they keepe the faith and order, as long as shee remains herselfe in purity of Doctrine and worship?

d Acts 8. 5. 14.
& 11. 22. 27. &
13. 2. 3.
e Mat. 23. 19.
Acts 13. 1, 2. 3.
& 8. 1.

4. How shall a Church propagate, and visit other Churches? shall they do it by their members, ordinary Christians, or by their Ministers, ᵈ or Pastors? shall they ᵉ intend such propagation, or stay, till by their numbers increasing, they are necessitated to swarme, or are persecuted abroad?

f Acts 1. 2, 3, 4.
& 2. 47. & 13. 1.

5. If by their Pastors, must not there bee more ᶠ Ministers then one in the first Church? how else can any be spared to goe abroad about such works upon occasion?

g Acts 1. 25, 26.
h Acts 6. 6. & 14.
23.

6. When they have planted other Churches, | must not the ᵍ first Church take care for the providing of Elders or Ministers for these new planted Churches, and ʰ ordain

them, and fometimes goe ^{i} or fend fome to teach them, *^{i} Act. 8. 14. & 11. 22.*
and uphold the worfhip of God among them?

7. How can any preach, unleffe he be ^{k} fent? and how *^{k} Rom. 10. 15.*
can he be fent, unleffe by impofition of ^{l} hands of the *^{l} Act. 6. 6. & 13. 3. 1 Tim. 4. 14*
Presbytery of the firft Church?

8. If fo, hath not the firft Church and the Minifters
thereof, Apoftlolical ^{m} power in thefe things? *^{m} Eph. 4. 11. Act. 1. 25. & 8.*

9. But have all ^{n} Churches and Minifters this power? *14. & 11. 22. 27. ^{n} Rom. 16. 1.*
are they able? have they learned men enough, to ^{o} water *^{o} 1 Cor. 3. 6.*
where they have planted? If fome fhould not be of the
^{p} *Quorum*, as it were, in ordinations, and the like, what *^{p} 2 Tim. 1. 6. 1 Tim. 4. 14*
order, peace, or unity can be expected? *compared.*

10. Againe, if all Churches and Minifters have this
power, equally, to exercife the work Apoftolicall; muft
they not all then goe, or fend abroad, to convert the Indians, and plant Churches? and how can all be fpared
abroad? Are all ^{q} Apoftles? all Euangelifts? where *^{q} 1 Cor. 12. 19, 29.*
were the body, if fo?

11. Will they not interfiere one upon another, and tref-
paffe upon one anothers' line, rule, or portion, which bleffed *^{r} 2 Cor. 10. 12. to the end.*
S. *Paul* condemned in thofe that entred into his labours?

12. When any other ^{f} Church, befides the ^{t} firft, hath *^{f} Act. 13. 1, 2, 3. ^{t} Act. 1. 4 & 2*
power and ability to propagate and bring forth other *47.*
Churches, may fhe not doe well fo to doe? muft fhe not?
in her fitting line, obferving peace, and holding commu-
nion with | the firft, as long as they remain in purity

both of them? and if a fecond, why not a third, and a fourth, and fo forth to a competent number?

13. Whether the firft and other Churches alfo having power and ability thus to propagate the Gofpell and plant Churches, may not be fitly called, prime, chief, or principall feats of the Church, or ^v chiefe Churches?

^v As Hierufalem, Antioch, Ephefus. Acts 11. 26.

14. Whether thofe Churches fo gathered, in one Kingdome, Citie, or Principality, holding communion together, may not be fitly, in regard of their unity in Doctrine and worfhip, called the Church of fuch a Nation, or Province, ^u City, or Countrey?

^u Acts 11. 22.

15. Whether is it probable, that the firft Church Chriftian, that wee reade of to be, at ^x *Hierufalem*, was onely one congregation, or but as many as could meete in one place? had they not among them twelve Apoftles, befides Elders, three thoufand, at once added, what ever number there was befides? and had they fuch a large Temple or meeting-houfes at their command in thofe dayes?

^x Acts 1. 4, 15, 26. & 2. 41.

16. Whether the word *Church* bee not diverfly taken in holy Scripture, and fometimes for a civill or uncivill affembly or congreffion? ^y *Acts* 19. 40. Καὶ ταῦτα εἰπὼν ἀπέλυσε τὴν ἐκκλησίαν, and when he had thus fpoken, he difmiffed the affembly or Church?

^y Act. 19. 40.

17. Whether anciently in *England*, fome fmall affem

Fitzherb. N.B.²²⁹

---

²²⁹ "For the word *Ecclefia* is always intended a parfonage." Fitzherbert's *Natura Brevium*, 32. "In a quare impedit præfentare ad Ecclefi-

blyes were not called Churches, as every | prefentative Rectory or Parfonage is called *Ecclefia*, when others that were greater were not fo called, as no Vicaridge, Donative or Chappel is called *Ecclefia* in our Law?

18. Whether the Rector, or Parfon that is a Presbyter in a Church, fhould, being alone, rule abfolutely by himfelfe, without the concurrence, advife, or fuperiour power of the Evangelifticall ᶻ Paftor of the Church, who had care in the plantation or erection of the Parfons Church?

19. If not; fhould the Vicar, Donative, Minifter or Chaplain?

20. But where they have ufed to rule more abfolutely, (as in fome peculiar jurifdictions in *England*) why may they not with the peace and unity of the Church, and by good advife, ftil doe the fame alway, with fubordination to the Evangelifticall leaders, and fit Chriftian, and Nationall Synods?

21. If the Parfon fhould not rule alone ordinarily, why fhould the principall leaders rule ordinarily alone without the advife and affiftance of a competent number of their Presbyters, who may afford them counfell? Did not the holy Apoftles advife with the Elders ᵃ fometimes? is it fafe for them or the whole?

22. But were there any Bifhops fuperintendent, over

am, it is a good plea to the writ that it is but a Chapel; for *Ecclefia* fhall be intended a parifh church." Lord Hale's *Comment. in loco*.

other Bishops, or Presbyters, in the first hundred years after Chrifts birth? Did not Saint *Iames* write his Generall Epiftle to the twelve | Tribes, which were then fcattered abroad, no doubt, in many places, and therein mention for Rulers, onely [b] Elders? and S. *Peter* write his generall Epiftle, and therein direct or command the [c] Elders not to over-rule the flock, the Lords inheritance? where was the Order of Bifhops? had not the Elders the rule? might they not elfe have returned anfwer, that the command concerned not them, but a certaine Order of men, called Bifhops, above us?

62

[b] James 5. 14.

[c] 1 Pet. 5. 1, 2. 3.

Anfwer. 23. Were not the Apoftles and Euangelifts then living, [d] Bifhops, and fuperintendent overfeers? had they not the [e] care of all the Churches, in their lines? did not thefe holy Apoftles, S. *Iames* and S. *Peter*, mention their owne names, in their Epiftles? is it not plain, that *Peter* had over-fight upon thofe to whom he wrote, to fee that they did not over-rule, and take account of them, if they did? And did the Lord ordaine there fhould be fuch a fuperintendencie, onely for an 80. years, and not fome equall correfpondent fuperfpection alfo in after-ages, when thofe extraordinary men fhould ceafe? If fome had then the care of all the Churches, fhould there not be fome, in after-times, to have the care of fome, to a competent number of Churches, in their fitting lines, and as they are [f] able? And though this Divine right be broken

[d] Acts 1. 20.

[e] 2 Cor. 11. 28. & Chap. 10. 12. to the end.

[f] 2 Cor. 8. 12.

through the many groffe corruptions of fucceffions, and the like, yet is it not equall to obferve the firft Inftitution, as neere as may be, as we fay the equity of fome Lawes and Statutes among us is fometimes to be obferved, though | not in the Letter? And why may not a chiefe 63 Paftor be called a Bifhop, as well as an Elder, or any other officer heretofore fuperiour?

24. If [g] Pfalms, and Hymnes, and fpirituall fongs are to be fung in the Church, and to fing melodioufly, and with good harmony, is the gift of God, and uncomely finging a kind of fin in the holy Affemblies; why fhould not the chiefe leaders, and rulers of the Church, appoint fome, in their ftead, to take care of the fingings of the Church? and may not fome be fitter to lead in finging, then others? and left they may fall out of their tunes to jarring, why may they not ufe the help of fome muficall inftruments? and left they fhould want able men this way, why fhould they not take care, that fome children be trained up in Mufique?

[g] Eph. 5. 19
1 Cor. 14. 26. 40

25. Whether or no Chrift did not allow of a [h] form of Prayer, and a fhort one too? will not the [i] ftrong allow the weak helps in Prayer? are not the beft Chriftians often diftracted in long Prayers? is it not eafier for the ftrong to pray, then for as ftrong men to hear Prayer well? fhould thofe that are ftrong Proficients in grace not be fatisfied, without all their weak brethren come to

[h] Mat. 6. 9. Sa ergo *moderate* τοι
οὕτως.
Luke 11. 2
[i] Rom 15. 1

k Rom. 12. 16.
Idipsum in invicem sentientes; non alta sapientes, sed humilibus consentientes. ἀλλὰ τοῖς ταπεινοῖς συναπαγόμενοι, but condescending to the humble.

the same pitch of high sanctification with themselves? should they not rather *k* condescend to the weaker? And although it be rare to tell of any actually converted by formes of Prayer, and Scriptures read; yet who can justly deny, but that much good hath been, is, and may for ever be done by such things that way, *Sicut ultimus ictus quercum non cædit, extrema arena clepsydram non exhaurit,* as the last stroak fells not the oake, nor the last sand exhausts the houre-glasse?[230]

l Act. 10 24, 44, 47, 48. & 16. 30. 31, 32, 33. versi. 14, 15.
m Acts 8. 8, 12, 14.

26. Whether may not a man *l* and his houshold, a woman and her houshold, a whole *m* City, or Countrey, a King and his people, a whole Nation, be baptized, after they are competently instructed in the Religion of God.

n Acts 10.
o Acts 16.
p Acts 8.
q Acts 18.

27. Is it certain, that all that were baptized in *n* *Cornelius* his house, in the *o* Gaolers house, in *Lydia's*, in *p* *Samaria*, in *q* *Corinth*, were such true beleevers, as now good men require all those that joyne with them, to be, before they will receive them to the Communion of their Church?

r Acts. 8. 13. Act. 2. 41. & 5. 1. compared.

Were not *r* hypocrites admitted & baptized in the Primitive Church, by the Apostles and Evangelists themselves, being deceived by them? Were not children circumcised in the old Testament, and baptized all along in the times of the New, so received into the bosome of the Church?

[230] "Quem admodum clepsydram non extremum stillicidium exhaurit, sed quidquid ante defluxit: sic ultima hora, qua esse definimus, non sola mortem facit, sed sola consummat." — Seneca, *Epist.* xxiv. 19.

28. Could, or can ever any Nation, probably, be brought into the obedience of the Gofpel, poll by poll, in fuch manner as is imagined by the leaders of feparations?

29. If it be poffible, let them make experience, and try whether the *Indians*, or any other Heathen people, can be fo converted before the Greek Kalends.

30. Whether there be any direct Scripture for the peoples choice of their chiefe Paftour? Can there, ordinarily, be a better election, then when the fupreame Magiftrate (who hath, at moft | times, the power of all the people, and fometimes their counfell in a regular way) joynes with a felect and competent number and company of Presbyters in the fame?

31. Whether any that have not fkill, grace, and learning, to judge of the parties to be ordained, whether they be fit, and able to what they are to be ordained, may ᶠ ordaine them?

ᶠ 1 Tim. 5. 22
Titus 2. 2.

32. Whether or no to maintain a defired purity or perfection in the Magiftracie, by election of the people, thefe good men of *New-England*, are not forced to be too ftrict in receiving the brethren, and to run a courfe tending to heathenifme?

33. Whether have not popular elections of chiefe Magiftrates beene, and are they not very dangerous to States and Kindomes? Are there not fome great myfteries of State and government? Is it poffible, conve-

nient, or neceffary, for all men to attain to the knowledge of thofe myfteries, or to have the like meafure of knowledge, faith, mercifulneffe, wifdome, courage, magnanimity, patience? Whence are Kings denominated, but from their fkill and knowledge to rule? whereto they are even born and educated, and by long experience, and faithfull Counfellors enabled, and the grace and bleffing of God upon all? Doe not the wife, good, ancient, and renowned Laws of *England* attribute much, yea, very much truft and confidence to the King, as to the head and fupreame Governour, though much be alfo in the reft of the great body, heart and hands, and feete, to counfell, maintain, and | preferve the whole, but efpecially the Head?

34. Hence what government for an Englifhman but an hereditary, fucceffive, *King,* ᵛ *the fon of Nobles,* well counfelled and affifted?

35. Whether we the pofterity of the Church, and people of God, who now fee the tops of things onely, may fafely condemne the foundations, which we have not feen?

36. Whether is there not a difference between bare fpeculation, and knowledge joyned with found experience, and betweene the experience of Divines and people reforming from out of fome deepe corruptions in Churches called Chriftian, and the experience of thofe that have converfed in and about planting, and building Churches,

ᵛ Pro. 25. 3.
Ecclef. 8. 4 &
10. 16, 17.

where there was none before, or among Heathens? what is art many times without experience?

37. Whether thofe Authors from *Hierome*,[231] to Arch-Bifhop *Adamfon*,[232] that alledge all Presbyters to be equall, and fhould alwayes have equall power and authority, had any great fkill, or will, or experience, in the propagation of Churches among heathens, or barbarous Nations?

38. If not, whether their Teftimony bee of that validity as is thought by fome? If they had, whether they might not erre?

39. Whether meffengers fent by Churches, or Minif-

---

[231] "Hæc propterea, ut oftenderemus apud veteres eofdem fuiffe prefbyteros et epifcopos. . . . Sicut ergo presbyteri fciunt fe ex ecclefiæ confuetudine ei, qui fibi propofitus fuerit effe fubjectos, ita epifcopi noverint fe magis confuetudine quam difpofitionis dominicæ veritate, presbyteris effe majores." Hieron. *Comment. in Tit.* i. 5 [cited, with other paffages from Jerome's Commentaries and Epiftles, in the Rev. Dr. Dexter's *Congregationalifm*, pp. 94–96; where fee a careful digeft of authorities, from Clement of Rome to Dean Alford, affirming the original equality of all presbyters.]

[232] Patrick Adamfon, titular Archbifhop of Saint Andrew's, Scotland, 1575–92, who had been a vigorous and uncompromifing opponent of Presbyterianifm, near the clofe of his life fubfcribed "certain articles allowing presbyterial difcipline and condemning the government epifcopal." "Whether he knew what was contained in them, or that he was induced thereto by a poor collection they gave him in the time (for fo the report went), or otherwife, it is uncertain," fays Bifhop Spottifwood. *Hift. of the Church of Scotland* (ed. Ruffel), ii. 415; comp. Calderwood's *True Hiftory*, 96; and *Stephen*, i. 299. Some years after the Archbifhop's death, thefe articles were printed, with the title of *The Recantation of Maifter Patrick Adamfon, fometime Archbifhop of S. Androwes in Scotlande*. (n. p. 1598.)

ters taking upon them to go to gather or plant Churches, and to ordain, or give the right hand of fellowship to Minifters in thofe Churches, | and to appeafe differences in Church affairs, are not Epifcopall acts?[233]

40. Is Epifcopacie, or a fuperintendencie neceffary at *New-England*, and is it not neceffary in more populous places? Are there not fome, nay many depths and [u] myfteries in Gods holy Word, the Scriptures, and certain Catholique interpretations, which tranfgreffed, the faith is hurt? Is it poffible, convenient, or neceffary for all men, nay all Minifters, to attain the knowledge of thofe myfteries, or to have the like meafure of knowledge, faith, mercifulneffe, wifdome, patience, long fuffering, courage, whereby to be enabled to rule in the Church of God, whereto they are educated, tryed, chofen, and ordained? and do not the facred rules and Laws of God, of holy Church and of this Kingdome attribute much, yea very much truft and confidence to the chief Paftors, Leaders, and Rulers, the Fathers of the Church, efpecially to the Bifhops of the prime and Metropoliticall Churches, by the affiftance of, and with, and under the fupreame Magiftrate, the chiefe, the beft cement of government, though much be alfo in other members of the great body, the Church, to counfell, maintaine and preferve the whole in the faith, foundneffe, peace and unity, efpecially the chief

[u] 1 Cor. 4. 1. & 2. 10.

[233] See before, pp. 53, 54.

Newes from New-England.

leaders, when need requireth? Hence what government for Chriſtians in chief, but by pious, learned, Provinciall and Dioceſan Biſhops, eſpecially in *England* and *Ireland?* By the juſt examination of the whole, thoſe that are pious and learned, may eaſily gather, what good | reaſons I had, and have, to returne, as now humbly I doe, to the Church of England, for whoſe peace, purity, and proſperity, is the daily prayer of one of her moſt unworthy ſons,

68

Clements Inne,
Novemb. 16. 1641.

Thomas Lechford.

To a friend.[234]

Sir,

HEre is a good Land, and yeelding many good commodities, eſpecially fiſh, and furs, corne, and other richer things, if well followed, and if that popular elections deſtroy us not. It is a good Land, I ſay, that inſtructs us to repentance, when we conſider what a good Land we came from, what good lawes and government we have left, to make experiments of governing our ſelves here by new wayes, wherein (like young Phyſitians) of

[234] A copy of this letter, in ſhort-hand, without date or addreſs, is in Lechford's Ms. Journal, pp. 164, 165. In the margin are the words (alſo in ſhort-hand), "This is written."

neceffity we muft hurt and fpoile one another a great while, before we come to fuch a fetled Common-wealth, or Church-government, as is in *England*.

I thank God, now I underftand by experience, that there is no fuch government for *Englifh* men, or any Nation, as a Monarchy; nor for Chriftians, as by a lawfull Minifterie, under godly Diocefan Bifhops, deducing their ftation and calling from Chrift and his Apoftles, in defcent or fucceffion; a thing of greater confequence then | ceremonies, (would to God I had known it fooner) which while I have in my place ftood for here thefe two years, and not agreeing to this new difcipline, impoffible to be executed, or long continued, what I have fuffered, many here can tell; I am kept from the Sacrament, and all place of preferment in the Common-wealth, and forced to get my living by writing petty things, which fcarce finds me bread; and therefore fometimes I look to planting of corne, but have not yet here an houfe of my owne to put my head in, or any ftock going: Whereupon I was determined to come back,[235] but by the over-entreaty of fome friends,[236] I here think to ftay a while longer, hoping that the Lord will fhortly give a good iffue to things both in our native Country, and *Scotland*, and here, as well as in all other his Majefties dominions.

[235] "To come back *into Ireland.*"— *Ms. copy.*

[236] "Of *my wife and* fome *other* friends."— *Idem.*

Newes from New-England.

I was very glad to fee my Lord Bifhop of *Exeters* Book;[237] it gave me much fatisfaction. If the people may make Minifters, or any Minifters make others without an Apoftolicall [238] Bifhop, what confufion will there be? If the whole Church, or every congregation, as our good men think, have the power of the keyes, how many Bifhops then fhall we have? If every Parifh or congregation be fo free and independent, as they terme it, what unity can we expect?

Glad alfo was I to fee Mafter *Balls* Book of the tryall of the grounds of Separation,[239] both which are newly come over, and I hope will work much good among us here?

And whereas I was fometimes mif-led by thofe of opinion that Bifhops,[240] and Presbyters, & all Minifters, are of

---

[237] "For Epifcopacie by divine right." *Idem.* Bifhop Hall's *Epifcopacie by Divine Right afferted*—a work undertaken at the requeft of Archbifhop Laud, and remodelled in conformity with his fuggeftions—was publifhed in 1640.

[238] "Apoftolicall *or Evangelicall* Bifhop." —*Ms. copy*.

[239] "A Friendly Trial of the Grounds tending to Separation : In a plain and modeft Difpute touching the lawfulnefs of a Stinted Liturgy and Set Form of Prayer ; Communion in Mixed Affemblies ; and, the Primitive Subject, and Firft Receptacle, of the Power of the Keys." &c. (1640, 4to, pp. 314.) See Hanbury's *Memorials*, ii. 46, 47, 156–63. In 1642, Mr. Cotton publifhed "A modeft and clear Anfwer to Mr. Ball's Difcourfe of Set Forms of Prayer," &c. John Ball, whom Fuller pronounces "an excellent fchoolman and fchoolmafter, a painful preacher, and a profitable writer," was minifter at Whitmore, near Newcaftle, in Staffordfhire. He died in 1640. *Worthies of England*, (ed. 1840) iii. 23 ; Brook's *Lives*, ii. 440.

[240] "Bifhops *diocefan were not of divine right and that Bifhops*, and Presbyters," &c. —*Ms. copy*.

the same authority; When I came to consider the necessary propagation of the truth, and government of the Church, by experimentall foot-steps here, I quickly saw my error: For besides, if the congregations be not united under one Diocesan in fit compasse, they are in a confusion, notwithstanding all their classicall pretendments, how can the Gospel be propagated to the Indians without an Apostolicall[241] Bishop? If any Church, or people, by the Kings leave, send forth Ministers to teach and instruct the poore Indians in the Christian Religion, they must have at least Apostolicall[242] power to ordain Ministers or Elders in every congregation among them; and when they have so done, they have power of Visitation where they plant: Nor can they without just cause[243] be thrust out from government without great impiety; and where they have planted, that is their line or Diocese. Thus I came to see, that of necessity a Diocese, and Bishop Diocesan, is very neere, if not altogether[244] of Divine authority.

I am also of opinion, that it were good for our Ministers to learne how to doe this work from some of our reverend Bishops in *England*, for I feare our Ministers

---

[241] "Apostolicall *or Evangelicall* Bishop." *Idem.* See before, pp. 59, 60, Queries 10–13.

[242] The Ms. copy has "*Evangelicall*" instead of "Apostolicall."

[243] The words "*without just cause*" are not in the Ms. copy.

[244] This qualifying clause, "very near, if not altogether," was inserted on revision.

know not how to goe about it. Whether muft not fome
Minifters learne their language? It is a copious language, as I am informed, and they have as many words
to expreffe one thing as we have. And when they teach
Indians to pray, will they not teach them | by a forme?
and how can Gods worfhip be maintained among ignorant perfons without a forme? I am firme of opinion,
that the beft of us have been much beholding to the
Word read, and formes of Prayer.

From Bofton *in* N. E.
Iulii 28. 1640.

This Gentleman[245] *to whom I wrote, kindly returned me a wife anfwer, wherein is this paffage:*

TO fpeak in briefe, I think now that *New-England* is
a perfect model and fampler of the ftate of us here
at this time; for all is out of joynt both in Church and
Common-wealth, and when it will be better, God knoweth: To him we muft pray for the amendment of it, and
that he will not lay on us the merits of our nationall and
particular finnes, the true caufe of all thefe evils.

Dated out of Somerfet-fhire,
Aprilis 27. 1641.

[245] William Prynne? He was a native of Somerfetfhire, and an old friend
of Lechford's. See the Introduction.
It was rumored in the summer of
1641 that he had fent money to Lechford to pay his paffage to England.

To another, thus: [246]

IN a word or two, we heare of great difturbances in our deare native Countrey; I am heartily forry, &c.[247] I befeech you take my briefe opinion; We here are quite out of the way of right government both in Church and Common-wealth, as I verily think, and as far as I can judge upon better confideration, and fome pains taken | in fearching after the bottome of fome things. Some electorie wayes tend to the overthrow of Kingdomes: No fuch way for government of Englifhmen, as a Monarchie; of Chriftians, as by Diocefan Bifhops[248] in their line: Better yeeld to many preffures in a Monarchie, then for fubjects to deftroy, and fpoile one another.[249] If I were worthy to advife a word, I fhould

---

[246] The draft of this letter (in fhorthand) is in Lechford's Journal, p. 175, with this note (alfo in fhort-hand) added: "This letter was fent by Mr. K. to his father, Ralfe King, of Watford." It is not certain (nor, I think, probable) that it was *addreffed* to Mr. King. It may have been fent to his care, to infure its fafe tranfmiffion to the perfon to whom it was written. The day before its date, Lechford had drawn, for Thomas Talmadge and his brothers, a letter of attorney to Ralfe King, of Watford, co. Herts, woollen draper, to receive for them certain moneys in England.

In the MS. the letter begins as follows: "Right Worthy Sir. I fent you at my firft landing here an unwife letter of which I [deferved ?] to receive no anfwer. I can not forget my refpect toward you and your worthy and beloved family, my good lady, and all your dear and hopeful children, as in my [poor?] fupplications I remember dayly. In a word," &c.

[247] "I am heartily forry that I had ever hand in fome of the *caufes.*" — *Ms.* [I think that I have not mifread the cipher, though the characters are fo imperfectly formed that I am not certain of the words italicized.]

[248] "Or Evangelifts." — *Ms.*

[249] "As I fear we muft do here

defire you to have a care, and fo all your friends, you prejudice not your eftate, or pofterity, by too much oppofing the Regall power: For I verily beleeve the Kings Majefty hath in generall a good caufe touching Epifcopacie:[250] My reafons I could better deliver in prefence, if haply God give opportunity to fee you, or if you require it hereafter, I will be ready to prefent my thoughts unto you. All this, as I fhall anfwer before the Lord, without any by-refpects. If you were here, I prefume you would fee more then I can, but I think you would be much of my mind.[251]

From Bofton *in* N. E. *Septemb.* 4. 1640.

To another of no meane rank.

COmplaining of my fufferings, and fhewing the rea- fons, defiring him to fend for me, that I might de- clare them to his perfon more effectually.

From Bofton *in* N. E. *March*, 1640.

long before we come to any fettled- nefs either in Church or Common- wealth." *Ms.*, — but this was croffed out on revifion.

[250] For "touching Epifcopacie," the Ms. reads, "againft the Sectaries."

[251] After this comes, in the copy, a paragraph about matters of bufinefs. "I hear that you [required?] that

20l. I owed you, of Mr. Hill. God's will be done. I am not able to pay it yet, but fhall be mindfull, God will- ing, to difcharge it as foon as I can. I am thankfull, and defire to be yet more thankfull to you for the loan of it. If you hear any thing of me fpecially from Mr. *Hooke* or his wife, pray keep an ear for me, for we have

73   *To another.*

YOu knew my condition and employment, and how ill it went with me in *England*, by reafon of the trouble of our friends, and my own danger therby. For my outward fubfiftence here, at this time, God knowes it is but meane; fome fay it is my owne fault, and that I ftand in my owne light, and you, and others may fo conceive; but the God of heaven is my witneffe, I have endeavoured in all things to keep a good confcience, though fometimes I have failed; I have endeavoured, laying all by-refpects afide, to joyne with the Church here, but cannot yet be fatisfyed in divers particulars, whereby I am kept from all place of employment or preferment, as I have had overtures made unto me of, if I would or could yeeld, but hitherto I have not dared to doe it, for good reafons beft knowne to our heavenly Witneffe. I muft give you a tafte.

They hold their Covenant conftitutes their Church, and that implyes, we that come to joyne with them, were not members of any true Church whence we came, and that I dare not profeffe. Againe, here is required fuch confeffions, and profeffions, both in private and publique,

had fome [*feveral words erafed*] and all yours to the guidance and . . .
I wifh you knew how I am ufed: of his heavenly Majefty, refting yours
For this time thus I take my leave in all fervice to be commanded.
heartily recommending your Worfhip   " THOMAS LECHFORD."

Newes from New-England.   15

both by men and women, before they be admitted, that three parts of the people of the Country remaine out of the Church, fo that in fhort time moft of the people will remaine unbaptized,[252] if this courfe hold, and is (we feare) of dangerous confequence, a thing not tending to the

[252] Robert Baylie, in *A Diffuafive from the Errours of the Time*, &c. (Lond. 1645), refers to this ftatement as his authority for the affertion that the fruits of the church-way of New England were, "firft, the holding-out of all their Churches and Chriftian Congregation many thoufands of People, who in former time have been reputed in *Old England* very good Chriftians." In the *Way of Congregational Churches Cleared* (the firft Part of which was written in reply to Baylie's *Diffuafive*), Mr. Cotton examines Lechford's teftimony: "The Book is unfitly called plaine dealing, which (in refpect of many paffages in it) might rather be called falfe and fradulent. I forbear to fpeak of the man himfelf, becaufe foon after the publifhing of that Book, himfelf was called away out of the world to give account [&c.] . . . That which he teftifieth, neither is it true; neither if it were, doeth it reach Mr. *Baylie's* affertion. It is not true, that three parts of the Countrey remaine out of the Church, if he meane three parts of foure, no, though hee fhould take in thofe remote *Englifh*, who live a fcore of miles or more from any Church." Pt. i. pp. 71, 72.

Right or wrong as to the *proportion* of non-members, Lechford was not the firft to complain of the ftrictnefs in admiffion to church privileges and of the virtual exclufion thereby of a confiderable, if not the greater, part of the people. Mr. Stansby, minifter of Little Waldingfield, co. Suffolk, in a letter to the Rev. John Wilfon, dated April 17, 1637, mentions as matter of grief, "that you [of Maffachufetts] are fo ftrict in admiffion of members to your church, that more then one halfe are out of your church in all your congregations, & that Mr. Hoker [Thomas Hooker] before he went away preached againft yt (as one reports who hard him) (& he faith) Now although I knowe all muft not be admytted, yet this may do much hurt," &c. "There is now," he adds, "fo much talke of yt, & fuch certeyne truth of yt, & I know many of worth, for outward eftate & ability, for wifdome & grace, are much danted from comeing." 4 *Mafs. Hift. Coll.*, vii. 11. Comp. Hooker, *Survey*, pt. 3, p. 6, — cited in note 12, p. 7, ante.

See, alfo, W. Rathband's *Briefe Narration of fome Church Courfes in New England*, (London, 1644.) pp. 9, 10.

74 propagation of the Gospel in peace: Which, though it have a colour of sanctimony and strictnesse, whereby many well-affected or affectionate people, but weak in sound experience and judgement, are the rather drawn thereunto, and they are in a manner necessitated to it, to maintaine their election of Magistrates and Ministers in their owne way of popular or Aristocraticall government; I dare not (for my part) yeeld unto neither in my own conscience, nor for the credit of those persons with whom I have been educated, and in whose causes I have been seen. A Monarchy is the best government for Englishmen; better to suffer some pressures under that kind of government, then to spoile one another with popular elections. Againe, I cannot yeeld to Lay-Elders, nor that Lay-men should impose hands upon any to the Ministerie, nor that any Minister should renounce his calling to the Ministerie which he received in *England*, as Antichristian: It is a grosse error, and palpable schisme; then our Baptisme is not right, and so there will be no end of separations. Also I beleeve there cannot be a Church, without a true Minister; nor can any gather themselves together into a Church without a true Minister; nor can they ordain their own Ministers; ordinarily, I meane; what may be done in an extraordinary case, *pro prima vice*, is another question; I hold there ought to be an Apostolicall Bishop, by succes-

sion from Christ and his Apostles, superiour in order or degree to his brethren; which Bishop ought to ordain, and rule with other Presbyters, or alone, but Presbyters cannot without him. And if so be any thing in word or act passed from mee to the contrary hereof, I do professe it was in my ignorance. Their calling is of Divine authority, or nearest thereunto, else the Church of God could not have subsisted in any tolerable way of peace, through all this by-past time of 1600. yeeres. I feare they know not what they say, that say the contrary: let them come here, they will quickly change their minde, if they study the point, and follow it home; for, besides the keeping of peace and unity, and a pure and able learned Ministery, how can the Gospell be propagated without some speciall Ministers, having the power Apostolicall, to goe forth to convert *Indians* or *Pagans*?[253] If a Pastor, or Minister, or Christian, of any Church shall doe so, what hath he to doe with Infidels? as hee is a Pastor, he is no Pastor to them. Therefore if any are sent to convert, and establish Churches among Infidels, such as are sent are Apostolick Messengers, Bishops or Ministers to them, and ought to be sent with fasting and Prayer, and by imposition of hands of the Presbytery, and having converted Infidels, may plant Churches, and ordain Ministers among them, and afterwards visit them; and is not this

---

[253] See before, pp. 21, 70; and Cotton's *Way Cleared*, pt. i. pp. 78, 79.

Epifcopacie, and their line wherein they have gone their Dioceffe? Thefe things naturally flow from, and are grounded in the Word, or equity thereof, and meere neceffity. Now if all Minifters fhould ordinarily have this authority, to go forth to thefe works, | without miffion, what quarelling there would be for divifion of Lines or Diocefes, let the experience of former ages tell, yea of the Apoftolique times, wherein were not wanting thofe that quarelled with Saint *Paul* himfelfe, about his Line or rule, 2 *Cor.* 10. Now unto this confufion, tends the opinion, that faith, a Bifhop and Presbyter is all one and equall; it is of *Acrius*,[254] it is falfe, and it is confufion. The reformed Churches and Writers that held fo, had little experience of miffion to convert & and plant Churches among Infidels. That reformation goes too deep that tends to pulling downe of Cathedrall Churches, and Bifhops houfes: Should not Apoftolick Bifhops, and the chiefeft Minifters have houfes to dwell in, and Churches

---

[254] For *Acrius*,—the name of a prefbyter of Sebafte, in Leffer Armenia, about the middle of the fourth century, who was the founder of a confiderable fect called Aerians. He taught that no difference ought to be recognized between a bifhop and a presbyter. He alfo condemned prayers for the dead, ftated fafts, and the celebration of Eafter.—Mofheim's *Eccl. Hift.*, bk. ii. pt. 2, ch. 3; Auguftine, *De Hæ-refibus*, c. liii.; Epiphanius, *Hærefis* 75 (ed. *Patav.*, pp. 905–912.)

Bifhop Hall, in *Epifcopacie by Divine Right afferted*, which Lechford had recently been reading (fee before, p. 69), mentions that "branded heretic Aerius," as "the only founder and abettor . . . in all the world of hiftory and record," of the opinions held by the difparagers of epifcopacy.—*Works* (ed. Wynter), ix. 246.

Newes from New-England.

to recide and officiate in, whither all the Churches of their Line may fend and come together in Councel, or Synod, and fo do nothing of great moment without their Bifhop, a *Timothy*, or a *Titus*? Again, Baptifme is admiffion and initiation into the Church; to whom Baptifme is committed, *viz.* Apoftles and Apoftolick Minifters, they have power of admiffion, that is, of loofing, and confequently of binding, excommunication or expulfion. Where is now the peoples power in the keyes? are they all Apoftles, and Apoftolick Minifters? what confufion is this? who can yeeld to it knowingly? I befeech you pardon my zeale, and when you have confidered all, pity my condition, and pray for me ftill. Well I am affured, that mafter *Prynne*[255] & mafter *Burton* would never yeeld to thefe things, efpecially, | if they had experience of them. It is good for us to fee our errours, and acknowledge them, that we may obtain peace in the day of account.

Bofton, 13. *Oct.* 1640.

To another.

SOrry and grieved we are at the heart, to heare of the troublous eftate and condition of our native countrey; wee here alfo meete with our troubles and diftreffes

---

[255] Refpecting Lechford's relations with William Prynne and his fellow-fufferers, Baftwick and Burton, fee the Introduction to this edition.

in outward things, and fome in fpirituall matters alfo.
Here wants a ftaple commodity to maintain cloathing to
the Colony. And for my own particular, hitherto I have
beene much diftreffed here by reafon I cannot yet fo
clearely underftand the Church proceedings, as to yeeld
to them, there are therein fo many difficult confidera-
tions, that they have fometimes bred great confufion in
my thoughts. Never fince I faw you have I received
the Sacrament of the Lords Supper. I have difputed in
writing, though to my great hinderance, in regard of outward things, yet bleffed be the Lord, to my better fatif-
faction at the laft. ' I never intended openly to oppofe
the godly here in any thing I thought they miftooke, but
I was lately taken at advantage, and brought before the
Magiftrates, before whom, giving a quiet and peaceable
anfwer, I was difmiffed with favour, and refpect promifed
me by fome of the chiefe for the future.[256] Our chiefe
difference was about the foundation of the Church and

[256] "I am fummoned to appear in Court to-morrow, being the 1ft of 10th, 1640. The Lord God direct me, &c."—*Short-hand note* in Lechford's *Journal*, p. 176.

"A Quarter Court held at Bofton the Firft Day of the 10th M°. 1640. ... Mr. Thomas Lechford, acknowl- edging hee had overfhot himfelfe, & is forry for it, promifing to attend his calling, & not to meddle wth contro- verfies, was difmiffed."—*Mafs. Col. Rec.*, i. 310.

Hon. James Bowdoin, introducing, in his Note to the Hiftorical Society's reprint of *Plaine dealing*, this ex- tract from the Records, remarked: "No allufion has ever been made to the caufe of [this decree]; but it feems to have been confidered as referring to the firft [of Sept. 3, 1639, by which Lechford was debarred from plead-

Newes from New-England.

Miniftery, and what rigid feparations | may tend unto, what is to be feared, in cafe the moft of the people here fhould remaine unbaptized; confiderations which may trouble the wifeft among us. Rigid feparations never did, nor can propagate the Gofpell of Chrift, they can do no good, they have done hurt. It is dangerous to found Church government on dark & uncertain interpretations of Propheticall, or other Scriptures; foundations ought to be full of evidence, & demonftration. Bleffed be the

ing]. The language, however, leads me to a different conclufion; but to what it does refer, I know not."— 3 *Mafs. Hift. Coll.*, iii. 400.

Lechford's *Journal* contains the draft (in fhort-hand) of his "quiet and peaceable anfwer." He ftates that he appeared before the Court in obedience to a warrant fent forth againft him, on an information by the Grand Jury, in September; but "fince that time (he fays) the General Court [of October 7th] was pleafed to fay fomething to me, when they brake up, as for good counfel to me, about fome tenents and difputations which I have held; advifing me to bear myfelf in filence and as became me. . . . According to that advice I have been hitherto, and fhall, God willing, be ready to carry myfelf for hereafter. . . . I defire not to trouble your Worfhips with long fpeech, to divert or hinder your other occafions; [but, waiving all the forms of trial and proof of the matters charged,] I defire your Worfhips to be pleafed to accept of this my fhort acknowledgment that I have, I do confefs, too far meddled in fome matters of church government and the like which I am not fufficient to underftand or declare; and although once I thought myfelf bound in confcience to fay fome of thofe things I have faid, yet now I am afhamed of many of them."

It will be obferved that the "fhort acknowledgment" is very adroitly framed. *What* things, formerly faid, he is now afhamed of, or *what* matters of church government he had unadvifedly meddled with, he leaves the court to conjecture. In letters to his friends at home, he was more explicit. See before, pp. 74, 75.

Perhaps the offence for which he was called to anfwer may have had fome connection with the queftions he propofed "to the Elders of Bofton," Sept. 9, 1640. See before, p. 55.

Lord, now fome of the chiefe leaders of the Churches here hold the Churches in *England* true Churches, and your Miniftery lawfull, though divers corruptions there may be among you;[257] yea fome there bee of the chiefe among us that conceive the government by godly Bifhops fuperintendent over others to be lawfull.[258] Churches are not perfect in this world. We may not for every difagreement in opinion, or for flender pretended corruptions, feparate from the Church: feparate fo once, and no end of feparation.

From Bofton *in* N. E.
Decem. 19. 1640.

To conclude.

Uppofe there are foure forts of Government, which are ufed in Church, as in Common-wealth; Monarchicall abfolute without Lawes, which is tyrannie;

---

[257] Baylie (in *A Diffuafive from the Errours of the Time*, &c.) quoted a private letter in which Mr. Cotton had declared that it was "an error, to conceive that our Congregations in England are none of them particular reformed Churches." "I willingly acknowledge," wrote Mr. Cotton, in his reply to Baylie's book, "I did appear againft that Error. But neither was I the firft that did appear againft it, (but divers godly Englifh Minifters before me:) neither have I fallen to the liking of the contrary opinion fince. But the Diffuader is much deceived, if he take that Error to be the judgment of the Churches of New-England, howfoever fome particular perfons may lean that way." *Way cleared*, pt. i. p. 18.

Comp. Welde's *Anfwer to W. R. his Narration*, pp. 45, 46, and 24.

[258] "Let no man think he [Lechford] was kept out of our Churches,

Monarchicall bounded by Lawes; Ariftocraticall, and Democraticall: Epifcopall abfolute, which is Popifh tyrannie; | Epifcopall regulated by juft Lawes; Presbyterian, and Congregationall: Which of thefe will all men like, and how long? Some have well compared the humour of the people in this kind, to a merry relation of an old man and his fonne, paffing through the ftreets of a City, with one horfe betweene them: Firft, the old man rode, then the people found fault with his unkindneffe, in that he did not caufe his fon to ride with him: then the young man gets up too, now the people fay they are both unmercifull to the beaft: downe comes the old man, then the young man is unmannerly to ride, and his father walk on foot: at laft downe goes the young man alfo, and leads the horfe, then they were both unwife to lead the horfe, and neither of them to ride. Well, but alter the inconftant vulgar will; if fo, God grant it be for the better. But then confider ftories, one alteration follows another; fome have altered fixe times, before they were fetled againe, and ever the people have paid for it both money and bloud.

Concerning Church-government, what the Presbyterian way is, and how futable for *Englands* Monarchie, I leave

for maintaining the authority of Bifhops. For we have in our Churches fome well refpected Brethren, who doe indifferently allow either Epifcopall, or *Presbyteriall, or Congregationall Government*, fo be it they governe according to the rules of the Gofpel." — Cotton, *Way cleared,* pt. i, p. 71.

to the pious experienced Divines to fet forth, and the Church and State thereof to judge.

And for the Congregationall independent government, whereof I have had fome experience, give me leave inftead of a better intelligencer thus to prefent to my deare countrey, now in a time of neede, my impartiall opinion in thefe confufed | papers: And in brief thus: Although it had fome fmall colour in Scripture, and a great pretence of holineffe, yet no found ground in the Scripture; Again, if it be neither fit nor poffible long to bee continued in *New-England*, as not I alone, but many more eye and eare witneffes doe know, and the learned can and will judge undoubtedly, it muft needes be much more unfit and impoffible to be brought into *England*, or *Ireland*, or any other populous Nation.

All which upon the whole I humbly fubmit unto the facred judgment and determination of holy Church, his royall Majefty, and his Highneffes great and honourable Councel, the high Court of PARLIAMENT.

Imprimatur,
Ioh: Hanfley.

FINIS.

INDEX.

INDEX.

ABORIGINES, their conversion neglected, 54, 55.
Adams, William, of Roxbury, 89.
Adamson, Patrick, 141.
Administration of the Lord's Supper, 45-48; of Baptism, 47, 48.
Admission of church-members, 18-29.
Advocates in court not allowed, 68.
Allen, Thomas, of Charlestown, 52, 82.
Allin, John, of Dedham, 83.
Alvord, Benedict, 88.
Angier, Sampson, 110.
Animals, domestic, 109; wild, 111.
Ann, Cape, 106, 112.
Apostles' creed, exceptions to some articles of the, 27.
Appeals to the king not allowed, 64.
Aspinwall, William, 64.
Assistants, how nominated, 60, 61.

BACON, Leonard, quoted, 33, 34.
Baptism, administration of, 47, 48.
Bastwick, John, his severe sentence, xvi.
Batchelor, Stephen, of Lynn, Yarmouth, and Hampton, 85.
Baylie, Robert, 54; his statements relative to the effect of New-England Congregationalism denied, 151.
Beggars rare, 69.

Bellingham, Richard, governor, 35, 86, 129.
Bells, what churches had them, 44.
Bilson, Bishop, his opinions on the descent of Christ into hell, 27.
Bishop, John, 91.
Bishops, diocesan, indispensable, 142, 144, 148.
Blackman, or Blakeman, Adam, of Stratford, 101.
Blackstone, William, 97.
Blackstone's Commentaries, quoted, 32.
Blackwood, Christopher, 93.
Blinman, Richard, 92, 107, 125, 126, 127.
"Body of Liberties" in 1641, its general character, 62.
Boteler, Lady Alice, wife of George Fenwick, 97, 98.
Bowdoin, James, xxxviii.; his remarks on a MS. copy of "Plain Dealing," xxxix.
Bracket, Richard, 86.
Bradstreet, Simon, 86, 125.
Braintree, a church formed there, 41.
Brinley, George, Dedication, 100.
Britton, or Brittaine, James, whipped for disrespect, 58, 94; hanged for adultery, 58.
Browne, Edmund, of Sudbury, xviii., 83.

203

Buckley, or Bulkley, Edward, of Marshfield, 126.
Bulkley, Peter, of Concord, his ordination, 16, 86.
Burdett, George, his misconduct, 105.
Burials, how conducted, 87, 88.
Burr, Jonathan, of Dorchester, 81.
Burton, Henry, his trial and severe sentence, xv.
Burton, Mr., xix.
Burton, Thomas, a petitioner with Robert Child, 82.

CAPAWACK, or Martha's Vineyard, 108.
Cape Ann, 106, 112.
Call of a Church, essential to ordination, 16, 17.
Cambridge Platform quoted, 32.
Catechising of children, 53, 54.
Charitable contributions, 48, 49.
Chancey, Charles, 89; his opinions on baptism, 90.
Cheever, Ezekiel, the father of New-England schoolmasters, 99.
Chickataubut, sachem of Neponset, 121.
Child, Robert, and others, their petition, 63, 64, 82.
Chirography, what it was in Lechford's time, xiv.
Christ's descent into hell, not necessary to be believed, 27.
Church, manner of gathering one, 12; church covenant, *ib.;* church officers, election of, 13; ordination of, *ib.;* church members, how received, 18–23; offending, how to be dealt with, 34, 35; no others can be freemen, 59; church censures, 30–34; churches not to be gathered without notice to magistrates and other churches, 73; how few may form a church, *ib.;* the Church, its relation to the State, 127; church government of New England disapproved by Lechford, 132–143; may a people form a church without a minister? *ib.;* "The Church, her Liberties," 72.
Civil franchise dependent on church-membership, 59.
Clapboards, primary meaning of, 111.
Clark, John, of Newport, 94.
Clement's Inn, xiii., xvii.
Climate, severe, 114.
Cobbett, Thomas, of Lynn, 84.
Cole, William, and wife Elizabeth, employ Lechford in a suit at law, xxvii.
Confession of faith, how made, 19–23.
Conversion of the natives, 54, 74, 77.
Cotton, John, of Boston, xxi., xxiv., xxxv., xxxvi., 81; his "Sermon of the Twelve Articles," 25; his Lectures on Revelation, 52, printed in London, *ib.;* his writings quoted, xxiv., xxxv., xxxvi., 13, 14, 17, 21, 27, 30–37, *et sæpe.*
Cotton manufacture, 110.
Court, General, meet semi-annually, 62; its powers, 63; place of meeting, 64.
Courts, Quarterly, 63.
Covenant, church, 12.

DALTON, Timothy, of Hampton, 85, 125.
D'Aulnay, his quarrel with La Tour, 108.
Davenport, John, of New Haven, 99.
Day, Stephen, the first printer, 123.

INDEX. 205

Days of the week and month, how designated, 54.
Deacons and deaconesses, their duties, 24.
Decline in prices, 113.
Denton, Richard, of Stamford, 97.
Diocesan bishops, needful, 142, 144, 148.
Douch, or Dutch, Osmond, 106.
Doughty, Francis, of Taunton, his difficulties, xxvii., 91, 92, 126.
Dover, 102, 103, 125.
Downing, Emanuel, xviii., 71.
Drums used to call people to public worship, 44.
Drunkenness rare, 69.
Dudley, Thomas, deputy-governor, his character, xxii.; Lechford consults him, *ib.;* his letter to Winthrop concerning Lechford's errors, *ib.;* mentioned, 86.
Dunster, Henry, of Cambridge, 82; commended, 122; his marriage, 123.

ELECTION of governor and magistrates, how conducted, 59.
Election-day, when, 61.
Eliot, John, of Roxbury, 81.
Endicot, John, 86.
England, laws of, not binding here, 62, 63.
Equality, original, of all presbyters, 141.
Episcopal ordination, how regarded, 16.
Excommunication, how pronounced, 30; a law concerning, 32.
Excommunicated persons, how treated, 31-34.
Exeter, 106.

FELLOWSHIP of the churches, how expressed, 14.
Fenwick, George, of Saybrook, 97, 98.
Firmin, Giles, of Ipswich, 84.
Fishing-business, 110.
Fisk, John, of Wenham, 84.
Flax, cultivation of, 110.
Flint, Henry, of Braintree, 41, 81.
Foote, Joshua, xxxvi.
Fordham, Robert, of Sudbury, 83.
Forms of Prayer, 137.
Fowle, Thomas, a petitioner with Robert Child and others, 110.
Freemen must first be church-members, 29, 59; their oath, 61.
Frost, William, 101.
Fuller, Samuel, 57.
Funerals, how conducted, 87.

GENERAL COURT, how often held, 62; place of meeting, 64.
George Ragotzki, or Rakoczy, Prince of Transylvania, xvii.
Gerrard, George, his letter to the Earl of Strafford, xii.
Gill, Thomas, 82.
Glover, Henry, an excommunicated man at New Haven, 34.
Glover, Josse, 123; his widow Elizabeth marries Henry Dunster, *ib.*
Gorges, Thomas, 104, 105.
Gorton, Samuel, 94, 95.
Governor and magistrates, how chosen, 59.
Grafton, Joseph, 107.
Grand juries, 64.
Green's Harbor, see Marshfield.
Grey, Henry, 101.
Grey, John, 101.
Grievances, 89.

INDEX.

HALLET, Andrew, 93.
Hartford, when it first had a bell, 44.
Heaton, Nathaniel, xix.
Hewes, Joshua, xxxvi.
Hewit, or Huit, Ephraim, 97.
Hibbins, William, of Boston, xxxvi.
Higginson, John, of Guilford and Salem, 98.
Hobart, Peter, of Hingham, 82.
Hooke, William, of Taunton, 90, 126.
Hooker, Richard, 80.
Hooker, Thomas, of Hartford, 97; quoted, 14, 17, 22, 30, 31, 39, 51, 57.
Humfrey, John, 86, 99, 114, 125.
Hurd, John, xix.
Hutchinson, Samuel, xix.

INDEPENDENCY of churches, 36.
Indians, their manners, character, habits, government, religion, &c., 115-122; they obtain fire-arms from the French and Dutch, 117; their powahes, or priests, 117; their condition improved from intercourse with the English, 117, 118; their religion, 120; names of their chiefs, 121; their conversion at present neglected, 54, 55; cannot be converted but by ministers episcopally sent, 153.
Inns of Chancery, why so called, xiv.
Inordinate church-going, 52.
Iron-works, 111.
Isle of Sable, 107, 108.
Isle of Shoals, 107, 110.

JACOB, Henry, a "Treatise" by him noticed, 27.
James, Thomas, of Charlestown, 99.

Jealousy of church power, 37.
Jenner, Thomas, of Roxbury, Weymouth, and Saco, 81, 105.
Jennison, Samuel, of Worcester, becomes the possessor of Lechford's MS. Journal, ix.
Jennison, Samuel, of Boston, x.
Jewel, John, Bishop of Salisbury, 80.
Jones, John, of Fairfield, 86; ordained at Concord, 16.
Josselyn, John, 77.
Juries, liberty of challenge restricted, 66.

KEAYNE, Robert, xix., xxiii., xxxiv., 126.
Keys, power of the, in the church, 30.
Kingly government preferred by the author, 140, 144, 148, 152.
Knight, William, of Ipswich and Topsfield, 84.
Knowles, John, of Watertown, 18, 28, 83.
Knowles, or Knollys, Hansard, 102, 103; his difficulty with Larkham, at Dover, 124.

LARKHAM, Thomas, of Dover, 103; quarrel with Knollys, 124.
Lashford, Sir Richard, xiii.
La Tour, his quarrel with D'Aulnay, 108.
Laud, William, Archbishop of Canterbury, xii., xv.
Laws of Moses followed, 65.
Lawyers held in small esteem, 68.
Lay ordination, 13.
Lechford arms, xiii.
Lechford, Sir Richard, xi., xii.; his daughters detained in England, xii.;

refuses the oath of allegiance, xii.; xiii. See Lashford.
LECHFORD, THOMAS, his MS. Journal, ix.; his family connections, xi.; account of him, xiii.; his account of himself, xiv., xv.; a lawyer, xiii., 4; a solicitor for Prynne in his trial before the Star Chamber, xv., xvii.; comes to Boston, xvii.; date of his arrival, xviii.; his wife, xviii., xix.; regarded in Boston with distrust, and why, xix., xx.; differs from the belief of the colonial churches, xx.; his alleged errors, xxi.; his letter to Hugh Peters, xvii., xxii., xxiv.; unsuccessful and disappointed in Boston, xxv.; his proposal to the General Court, xxv., xxvi.; his autograph, xxvi.; employed in the case of William Cole against Francis Doughty, xxvii. (see Doughty); his indiscretion, *ib.;* censured by the court, *ib.;* his confession of his fault, xxviii.; obtains employment from the magistrates. xxix.; counsels submission to the king, xxx.; dislikes popular government both in Church and State, xxxi.; becomes obnoxious to the magistrates, *ib.;* yet treated with remarkable lenity, xxxii.; summoned before the court, xxxiii.; censured, xxxiii., 156; submits, xxxiii., 157; implicated in the famous "sow case," xxxv.; returns to England, xvii., xxxvi., 109; his death and character, xxxvii.; value of "Plain Dealing," xxxviii.; asks forgiveness of the reader for his acts against Episcopacy, 3; his reasons for printing "Plain Dealing," 3, 4; his objections to Independency, 5; how long a resident in New England, 11; date of his departure from Boston, 35; perhaps took notes of Cotton's Lectures, printed in London, 52; makes sundry copies of the colonial laws for Gov. Winthrop and others, 65, 72; his "Propositions to the General Court," 69, 70; not employed by the court, and why, 71; his paper of advice to Gov. Winthrop, 76–80; extent of his travels in New England, 114, 115; several things which he disapproved, 129–131; discontented, 144, 150; commends Bishop Hall's book, "Episcopacie by Divine Right asserted," 145; eschews republican government, 148; a decided monarchist, 140, 144, 148, 152; believes that a church cannot exist without a minister episcopally ordained, 152; holds that bishops are the successors of the apostles, xxi., 152; thinks the Indians cannot be converted without episcopal authority and ordination, 153; regards congregationalism as unfit both for Old England and New, 160.

Lectures, public, 51, 52.

Lenthall, Robert, of Weymouth and Newport, xxiii., 17, 57, 94.

Leverich, or Leveridge, William, of Dover, Duxbury, and Sandwich, 92.

Limits of civil and ecclesiastical authority defined, 36.

Linen manufacture, 110.

Lions, their cry supposed to be heard, 112.

Long Island, 101; colonized from Lynn and Ipswich, 102.

INDEX.

Lord's Supper, manner of administering, 45, 46.
Lothrop, John, of Scituate and Barnstable, 93.
Loveran, John, 83.
Ludlow, Roger, 100.

MAGISTRATES, their power, 35; how chosen, 59.
Maine, province of, 105.
Majority, shall it rule? 38, 39.
Manufactures, 109; encouragement of, 110.
Marblehead, 40; incorporated, 41.
Marriage performed by the civil magistrate, 86, 87.
Marshal, an officer of the law, 67.
Marshall, Thomas, xix.
Marshfield, 92, 107, 125.
Martha's Vineyard, 108.
Martin, Ambrose, dislikes church covenants, 57.
Massasoit, 121.
Mather, Cotton, his opinion respecting public confession, 21.
Mather, Richard, 81, 126.
Matthews, Marmaduke, 93.
Maverick, Samuel, 106.
Mayhew, Thomas, 108.
Mayo, John, of Barnstable, 93.
Meeting-house, first, in Boston, described, 43.
Members of a church, no others can be freemen, 59.
Michelson, Edward, of Cambridge, 67.
Micklethwaite, Nathaniel, xviii., xix.
Military trainings, 89.
Miller, John, of Rowley, 84.
Milward, or Millard, Thomas, of Cape Ann, 106.

Ministers in the colony, list of, 81; how supported, 50, 51; their meetings, 37.
Minority of a church put under censure, a contrivance for excluding their vote, 39.
Monarchy the best government, 144, 148, 152.
Money, scarcity of, 113.
Moody, Lady Deborah, buys John Humfrey's farm, 98, 99.
Morality, strict, of New England, 69.
Mount Wollaston, 41.
Music in churches, the subject discussed by Lechford, 137.

NEWFOUNDLAND, Lechford goes to England by way of, 109.
Newman, Samuel, of Weymouth and Rehoboth, 81.
Newton, Joan, 88.
Nocake, what? 119.
No-man's Land, 108.
Nomination of Assistants, 60.
Norris, Edward, of Salem, 84.
Northam [Dover], 102, 103, 125.
Norton, John, of Ipswich and Boston, 84, 90.
Notary Public, the office of, proposed by Lechford, 70.
Nowell, Increase, 86.
Noyes, James, of Newbury, 56; his liberal views, 85.

OBJECTIONS to congregationalism stated, 132-143.
Offences, how heard, 29, 30.
Offley, David, xxxvi.
Oliver, John, instructs the servants at Rumney Marsh, 40.
Oliver, Thomas, 129.

INDEX.

Ordination, how performed, 13; episcopal, how regarded, 16; does not confer an indelible ministerial character, 17; may it be performed by laymen? 128, 129.
Original equality of all presbyters, 141.
Otis, John, of Hingham, 82.

PANTHERS, mistaken for lions, 112.
Parker, James, of Portsmouth, 81.
Parker, John, of Taunton, 91.
Parker, Robert, a learned non-conformist in England, 79, 80, 85.
Parker, Thomas, of Newbury, 56, 85.
Parker, William, 91.
Parties in court plead their own cause, 69.
Partridge, Ralph, of Duxbury, 93.
Pastors and teachers, distinct offices, 17, 18; debate on this subject in the Westminster Assembly, 18; shall they be chosen by the people? 139.
Patent of the colony demanded back, 65, 76, 77.
Peck, Robert, of Hingham, 82.
Penn, James, appointed beadle, or marshal, 67.
Pequonnock, 101.
Peters, Hugh, xiv., xvii., xxi., xxiv., xxxvi., 84, 102, 104, 111, 124; his testimony to the strict morality of New England, 69.
Philip, John, of Dedham, 83.
Phillips, George, of Watertown, 16, 18, 83.
Pierson, Abraham, of Southampton, 101.
Popular elections dangerous, 139, 152.

Powahes, Indian priests, or conjurors, 117.
Printing introduced, 123.
Productions of the soil, 109.
Profane swearing scarcely known, 69.
Profession of religion, how made, 18-23.
Prophesying, what? 42, 43.
Providence Island, one of the Bahamas, xvii., 113.
Proxy, voting by, allowed, 60.
Prudden, Peter, 97, 100.
Prynne, William, his trial in the Star Chamber, and severe sentence, xv., xvi.
Psalms, version of, used, 45.
Public lectures, 51, 52.
Public worship, 43.

QUERIES concerning Church Government, 132-143.

RAGOTZKI, or Rakoczy, George, Prince of Transylvania, xvii.
Randall, Abraham, 88.
Rashley, Thomas, of Cape Ann, 106, 107.
Rathband, William, his statements respecting the colonial churches, 17, 39.
Rattlesnakes, 112.
Records of proceedings in court sparingly made, 67.
Reyner, John, of Plymouth, 89.
Richmond's Island, 107.
Right hand of fellowship, 14.
Rogers, Ezekiel, of Rowley, 43, 84.
Rogers, Nathaniel, of Ipswich, 84.
Rowley, its early manufactures, 110.
Ruling elders, 24.
Rumney Marsh, now Chelsea, 40.

INDEX.

SALTONSTALL, Richard, a magistrate, 86.
Savage, James, xviii., xx., xxxiii.
Savage, Thomas, xix., xxxiv., xxxv.
Saxton, Peter, of Scituate, 93.
Scottow, Joshua, of Boston, 44, 52.
Shepard, Thomas, of Cambridge, 82.
Sherman, Mary, xix.
Sherman, Richard, of Boston, xxxiv.
Ship-building, 110; stimulated by Hugh Peters, 111.
Skelton, Samuel, of Salem, 37, 42.
Smith, Henry, of Wethersfield, 97.
Smith, John, fined for promoting the election of Robert Lenthall as minister of Weymouth, 58.
Smith, Ralph, of Plymouth, 90.
Snakeweed, an antidote to the poison of the rattlesnake, 112.
Snows, deep, 114.
"Sow case," xxxv.
Squire, John, 110.
Squire, Nicholas, 110.
Star Chamber, xv.
Stoddard, Anthony, of Boston, 35.
Stone, Samuel, of Hartford, 97.
Story, George, xix., xxxiv.
Stoughton, Israel, xxiii., 31, 86, 100.
Strafford, Earl of, xv.
Strawberry Bank [Portsmouth] patent, 125.
Street, Nicholas, of Taunton and New Haven, 90, 91, 126.
Strictness of the colonial churches, 151.
Support of ministers, 50, 51.
Symmes, Zechariah, of Charlestown, 82, 99.
Symonds, Henry, 102.
Symonds, Samuel, of Ipswich, register, 71, 125.

TAXATION for support of ministers, sometimes resorted to, 51.
Thomas, William, of Marshfield, 93, 125, 126.
Thompson, Maurice, a merchant of London, concerned in fishing at Cape Ann, 106.
Tomlyns, Edward, 102.
Tomlyns, Timothy, 102.
Tompson, William, of Braintree, 41, 81.
Trelawney, Mr., 107.
Trials, how conducted, 66.
"Twelve Articles of Religion, The," a sermon by John Cotton, 25.

UNDERHILL, Capt. John, 103, 104, 124; his difficulties at Dover, 104.

VEGETABLE productions of Massachusetts, 109.
Verdicts of juries, sometimes at random, 67.
Vines, Richard, of Saco, 105.
Voting for governor, &c., how conducted, 60; by proxy allowed, *ib.*

WARD, John, of Ipswich and Haverhill, 84.
Ward, Nathaniel, of Ipswich, 84; frames a body of laws, 64, 65; his advice to the General Court, 68; his testimony to the strict morality of New England, 69.
Ware, Mary, 88.
Warham, John, of Windsor, 97; his views touching admission to church privileges, 57.
Watertown, its church-bell, 44.
Watson, John, 101.

Webb, "alias Evered," John, 107.
Weld, Thomas, of Roxbury, xxvi., 13, 81; quoted, 15, 17, 23, 29, *et sæpe*.
Wentworth, Thomas, Earl of Strafford, xv.
Weston, Francis, of Salem, 66.
Weymouth Church, 15, 16, 17.
Wheelwright, John, 106.
Whitefield, Henry, of Guilford, 98, 100.
Whiting, Samuel, of Lynn, 84.
Wiggin, or Wiggon, Thomas, 125.

Williams, Roger, 42, 96; quoted, 17, 27, 32, 38, *et sæpe*.
Willis, John, xl.
Wilson, John, of Boston, 16, 81, 125, 126.
Winslow, Edward, 126.
Winter, John, 107.
Winthrop, John, xx.; quoted, 14, 16, 18, 31, 35, 36, *et sæpe*.
Winthrop, Stephen, recorder, 71, 86.
Woollen manufacture, 110.
Worcester, William, of Salisbury, 85.
Worship, public, how conducted, 44.

www.ingramcontent.com/pod-product-compliance
Lightning Source LLC
Chambersburg PA
CBHW021729220426
43662CB00008B/770